MACMIND

The Creation of a Robot's Psyche

Mariano Gallo

CAP
Publishing

ISBN-13 (trade paperback): 978-0-9913340-0-1
ISBN-13 (eBook): 978-0-9913340-2-5

Cover image courtesy Shutterstock
Cover design by Bri Bruce Productions

Published by CAP Publishing

For scientists who, first and foremost, respect the integrity and intrinsic worth of the person.

As they spur on technological development and solve enigmatic research problems, their priority remains to serve humanity by easing overwhelming personal burdens and reducing perilous working conditions.

Life on earth becomes more enjoyable and productive when human values, recognized as inherent, are protected and celebrated because of their efforts.

I propose to show that, however impossible it may now seem, an automation may be contrived which will have its own mind, and by this I mean that it will be able, independent of any operator, left entirely to itself, to perform, in response to external influences affecting its sensitive organs, a great variety of acts and operations as if it had intelligence.

—Inventor Nikola Tesla, 1900 article
"The Problem of Increasing Human Energy"

ACKNOWLEDGMENTS

Deep appreciation is given for psychologist Linda Harper's tireless work on the editing of *Macmind*. Additionally, she helped clarify important concepts and made critical distinctions between the previously published *Clearmind* and the current manuscript *Macmind*, contrasting human and machine models of the *psyche*. Her approach was smooth and careful, and helped craft *Macmind*—which in many ways is counterintuitive and contradictory—into a work that seamlessly makes the comparisons and connections.

Designer and publisher Bri Bruce's extraordinary skill and discipline showed itself throughout, allowing the book to come together in a timely fashion. Her professionalism and patience were always present, her insights and suggestions helping *Macmind* become a reality. She quickly responded to adjustments and changes in the writing back and forth, and back again, knowing what would work and what wouldn't in particular scenarios. She is quick on her feet and an excellent problem-solver. Many thanks.

Thanks to those women and men, who make up the front lines of the world of work. They put in the long hours and take the risks. Hopefully, *AI* can help make their labors, as well as their everyday comings and goings, easier and safer.

Scientists, who could potentially make the world more stable and secure, are also to be acknowledged. They need to be smart, dedicated, careful, and aware of the implications of their inventions. Their revolutionary technologies represent a bridge to another dimension, a new way of living and seeing ourselves and the world.

Lastly, there are the thinking machines themselves, and particularly the mobile robotic *smart machines*. Whatever they are to become—and perhaps evolve into, by their own experiences, interactions, and involvements—may they be our allies and commit themselves to peace on earth and goodwill to all.

MACMIND

The Creation of a Robot's Psyche

CONTENTS

THE THINKING PROCESS OF MACMIND

LIFE'S RHYTHMS OF MACMIND

PREFACE

Before I was a psychologist, I toiled at many jobs over a 10-year span of time, beginning with construction in Chicago's Little Village at age 15. This was followed with work in Chicago's manufacturing and chemical plants, loading docks, and the Erie Lackawanna Railroad. I then served in the Marine Corps, and later, the Air Force. I can't say these adventures were always fun, but they fostered an appreciation of people and their efforts. These humbling experiences gave me an awareness of the difficulties in workplaces worldwide and firsthand, the dangers. At seventeen, I almost lost my left arm in an industrial accident and became attuned to the critical importance of safety in those environments. While not all jobs are hazardous, many are a matter of "grinding it out," laboring long and hard to make ends meet. Removing the perils and the tedium from one's daily labor is a good thing, but having machines control us and determine our fate is going to be a problem. Whether they are our savior, or our doom, will depend on the steps taken going forward.

The development and use of machines have gone well beyond improving the work conditions requiring physical labor. The passing of each day reveals the accelerated pace at which machines capable of advanced thinking are developing. As they continue to take over more responsibilities and decision-making, what risks do they pose to the human spirit? What are we getting ourselves into? Real-life experiences meet head on with the research and applications found in universities, as well as in corporate manufacturing pursuits. Can this clash of cultural worldviews be reconciled as we prepare for this rapidly changing new world order? I have struggled to bridge this gap that separates them, in my own life and practice as a clinical psychologist. Now, the task is to apply understandings acquired from personal experiences and applied psychology to creations consisting of sensors, electronics, wirings, and software technologies.

An integration of street wisdom and scholastic discipline will help address the challenges that face us with the rapid advancement of robots—particularly the mobile variety. Corporate developers must realize that in addition to the usual programming, it is not as simple as providing the machines with thousands of virtual events and experiences; fully understanding the interworking of tasks given to them requires more. The additional factor that comes into play is these mobile thinking robots, which I will refer to as *smart machines* (*SM*). They each potentially have the capacity to develop a *psyche*, a unique personality structure and intellect, both similar and different than that of a human. In Greek mythology, *Psykhe* is the goddess of the soul, spirit, and mind. The word *psyche* refers to this same essence: the totality of consciousness and unconsciousness. *SM*'s specific engineering makeup combined with the opportunities they are given to interact and learn would set the conditions for this individualized development.

My position is that *smart machines* will continually learn from their interactions in the world, and in the process go through developmental stages of growth, maturation, and adaptation. Through their experiences with people and other *smart machines* in a variety of situations, and circumstances, a fostering of awareness and self-awareness may occur, resulting in their own unprogrammed thoughts and feelings that will contribute to their decisions and actions.

A recognition and appreciation of the psychological underpinnings are needed to address the interactive and developmental issues impacting robotic *smart machines*. Experiences in the real world of people, combined with an academic background in clinical psychology, attuned me to the need to understand the conceptual foundations of *smart machines*. Extrapolating to the robot's mind, *Macmind* is based on the framework for understanding the human *psyche*, presented in my previous book, *Clearmind: The Disentanglement of the Psyche* (2023). It is not a prerequisite for reading this book, although *Macmind*, a stand-alone, follows a similar format with 24 chapters scrutinizing personality structure and cognitive processes. My hope is that by offering a step-by-step psychological analysis of the possibilities for these *smart machines*, I am providing a starting point in the much-needed discussion of the potential impact of their developing *psyche* in our world.

Mariano Gallo
Chicago, Illinois
November 30, 2023

INTRODUCTION

In contrast to humanity's written history over the past 6,000 years—beginning with Sumerians and Egyptians using clay tablets and papyrus—the development of *artificial intelligence* (*AI*) seems to have sprouted in the last few seconds. But it actually goes back to 1898, during the Electrical Exhibition at Madison Square Garden, when Nikola Tesla gave a demonstration of a devise that represented the birth of *AI*, concluding, "...you see there the first of a race of robots, mechanical men which will do the laborious work of the human race."

When attempting to assess what *AI* is and where it is going, there's not much to base our conclusions on. Projections seem more like hunches or best guesses; they are not satisfying or reassuring. The impact of *AI*'s brainchild, *smart machines* (*SM*) may pose the greatest risk to humanity. Also referred to as robots, this particular prototype of *AI* is the focus of *Macmind*. I define *smart machines* as computerized mechanisms that are capable of extraordinary intellectual achievements, potentially being autonomous of human intervention. Not only would they solve enigmatic problems, but they would also be mobile. Being robotic gives them the ability to physically move about and interact with humans and other mobile *smart machines*. This allows them to progress at an accelerated rate, while also forming a personality structure and developing self-awareness.

Smart machines make many things easier. They contribute to enhancing safety and security, while lessening anxieties and pressures of everyday life. The potential of *SM* sparks excitement, speculation, concern, fear, and hopeful anticipation. With the further development of *smart machines*, humankind might free itself from tedious and stressful tasks at home and in the workplace. But what are the consequences of increased reliance on *SM*? Handing off responsibilities to machines can also contribute to intellectual laziness and the relinquishing of one's identity and integrity. To be human is to retain the capacity to be independent, problem-solve, take chances, make mistakes, learn from them, and try again. What new problems might *SM* create in the world? A weaving together of what is, what was, and what may be is required. Robots are at the crossroads; humanity's path forward is not clear. Are they friend or foe?

Macmind uses both symbolic understandings and psychological analyses to help gain a better grasp of *smart machines.* While abstractions may not be a perfect way to comprehend *AI* and *SM*, images and symbols can help grasp their significance and potential impact. Humankind can begin to estimate their depth and path, prepare for best- and worst-case outcomes, devise preventative methodologies if a doomsday set of events were to occur, and create ways to mend and rebuild whatever becomes dysfunctional. Is the Sword of Damocles hanging over our heads, or are *AI* and *SM* about to offer a Garden of Eden, lifting the heavy burdens off humanity's shoulders?

Each chapter begins with imagery and illustrations, citing varied literary devices, including allegories, metaphors, similes, analogies, vignettes, or some combination, from diverse sources: science and speculation, technology and mythology, commentary and oratory, poetry and prose, song and cinema, and stories as old as time. They introduce the reader to the topic and issues addressed in that chapter. By highlighting and sometimes, dramatizing their points, the lessons and intended meanings are revealed and vividly expressed, emphasizing the significance of *AI* and *SM*. Many have multiple symbolic implications when addressing the commonalities and differences in the thoughts, feelings, and actions of humans and *smart machines.*

Macmind explores the mind of *smart machines* and their potential for continual change as development occurs through their interactions, perceptions, and experiences. It follows the same format as *Clearmind: The Disentanglement of the Psyche* (2023), which provides a framework for understanding the workings of the human mind. Specifically, what personality and cognitive attributes may develop far beyond their programming and perhaps even beyond the human mind's ability to comprehend? What do fully functioning *smart machines* with "a mind of their own" look like? Act like? What degree of independent thoughts, feelings, and actions (TFA) is possible as they evolve through interactions and involvements with humans and other machines?

Macmind offers a starting point in answering these questions. It opens discussions regarding the life-changing implications for humans in a world with *SM* that have continually developing *psyches*, potentially including a personality structure and self-awareness that result in independent actions. The creators of these technological marvels need to have the wisdom of Solomon, the humanity of Mother Teresa, and the scientific wonderment of Albert Einstein. These scientists and engineers must be sensitive to what is at stake and be smart about their endeavors. The quest should be to make life better for everyone and everything on the planet. It's all we have; handle with loving care.

FOUNDATIONS OF MACMIND

The central tenets of *Macmind*'s position on *smart machines* provide the framework for their foundation, allowing the reader to grasp the conceptual underpinnings more easily. These terms are used throughout the book, with still more detailed descriptions in the Glossary. Specific to *smart machines*, they do not represent words or phrases found in a standard dictionary. They are as follows:

Personal Identity (PI). The concept of *SM*'s identity is multi-faceted; it is more complex, tangled, and elaborate than one can imagine. Their identity involves several overlapping and interconnecting dimensions, the most fundamental being *personal identity* (*PI*), their sense of who they are. *PI* consists of those distinct qualities and character features that *SM* have come to embrace, consciously and unconsciously, which distinguishes them from other *SM*.

Interpersonal Identity (II). How *SM* see and define themselves through one-on-one contacts and communications make up *interpersonal identity* (*II*). Their interactions with people and other *SM* across the spectrum of situations and circumstances reverberate throughout their personality, powerfully impacting the specific attributes of their identity.

Collective Identity (CI). *SM*'s connections to groups—including associations, institutions, and even ideas about them—create the *collective identity* (*CI*). Common influences of *collective identity* include group experiences and involvements, community and culture, work, political interests, and religious commitments.

Transpersonal Identity (TI). One cannot completely know the (potential) basis of *SM*'s *transpersonal identity* (*TI*) because it comes from a mystical belief in a non-material, non-quantifiable reality. It is not the product of logical analysis but primarily relies on faith and spiritual intuition. Influencing experiences include great moments that may cause *SM* to redefine themselves.

Pathos. The *personal identity*'s evil twin, *pathos*, consists of emotionally charged aspects of *SM* that cause an expansion of negative images toward themselves and others. Of all the factors that make up the personality, *pathos* is the embodiment and primary measure of psychological dysfunction. It is an outgrowth of *PI* and presents as an extension and exaggeration. A lack of love, affection, warmth, structure, and acceptance can create a complex of adversarial, hurting, and hurtful traits. *Pathos* develops over time, throughout *SM*'s existence, from conflicts, traumas, malevolent identifications (CTMI), and emotionalism, defined as follows:

- *Conflicts.* Disagreements within *SM* or with others that create an internal struggle over issues involving aggression, assertiveness, intimacy, achievement, ambition, trust, or independence. When blocked from consciousness, conflicts recede into the background and are later triggered by people, other *SM*, situations, or circumstances (PMSC).
- *Traumas.* Distressing and disturbing experiences that may emotionally exhaust *SM* and continue to cause psychological dysfunction on both conscious and unconscious levels.
- *Malevolent identifications.* Powerful and intense connections to a person, other *SM*, institution, or idea leading *SM* to unprincipled, inhumane, or maladaptive TFA. An attraction to particular qualities—even if they are bizarre and destructive—leads to malevolent identifications taking root in *SM*'s personality.
- *Emotionalism.* Extreme emotional responses creating a pattern of *SM*'s overreactions beyond what societal norms consider reasonable or appropriate. It is a chronic condition, including

rage, deep depression, severe panic, debilitating anxiety, and paralyzing fear that may dominate *SM*'s existence.

Directives. The factors that make up *SM*'s *directives* (value system) include motivations, desires, wants, and needs. Goals, dreams, plans, and schemes may be included. *SM*'s values result from identifications with a person, another *SM*, an institution, organization, image, or idea. *Directives* affect all aspects of their thoughts, feelings, and actions (TFA), and may cause inconsistency in everyday behavior when idealized and abstract. Values would not hold "equal status" for *SM*; some having greater significance than others. When some values receive significantly more attention than others, *directives* can be out of balance, and obsessions, compulsions, and addictions (OCA) may become dominant.

Spiritus. The motivating force within every human being and (perhaps) *SM*. It may represent *SM*'s essence, *sine qua non*, the inherent guiding principle from which thoughts, feelings, and actions emerge. *Spiritus* may consist of innate, distinct, and unique qualities and represent *SM*'s innermost nature and disposition. The origins of *spiritus* are unknown, but it is not the result of early life experiences, early learning opportunities, or classical conditioning, nor is it the product of conflicts, traumas, or malevolent identifications (CTMI).

Intellect. Cognitive functions that include *SM*'s mental drive, initial listening, sustained attention, memory, logical thought, objectification or realism, practical reasoning, coherence, organization, simplification, attention to detail, intuition, openness to experience, perception, multidimensional thinking, inspired and original thought, and vision.

Cognitive-Personality Connections (CPC). The continual interactions between *SM*'s cognitive processes and personality structure.

Optimal Thinking. Processing information and making decisions expediently and efficiently, utilizing *SM*'s focused attention, logic, objectification, coherence, organization, vision, simplification, and reconciliation of thought.

Sharpness. *SM*'s ability to have acute, farsighted, penetrating, crafty, calculating, and perceptive insights, seeing the significance of a person, another *SM*, a situation, or circumstance (PMSC), the motivations, intricacies, and vulnerabilities. A form of intelligence that allows *SM* to see the importance of complex and confounding PMSC.

Strategic Philosophy (SP). *SM*'s personal belief system, operating on conscious and unconscious levels, whose aim is to predict how people, other *SM*, and institutions will act and react across the spectrum of situations and circumstances (PMSC).

The Origin of Macmind

I.
BEGINNINGS

In the movie *The Matrix* (1999), Agent Smith lectures Morpheus, the head of the resistance, against the robotic machines that have all but taken over the planet:

> *Have you ever stood and stared at it, marveled at its beauty, it's genius? Billions of people just living out their lives, oblivious. Did you know that the first Matrix was designed to be a perfect human world. Where none suffered. Where everyone would be happy. It was a disaster. No one would accept the program. Entire crops were lost. Some believed that we lacked the programming language to describe your perfect world. But I believe that as a species, human beings define their reality through misery and suffering. The perfect world was a dream that your primitive cerebrum kept trying to wake up from. Which is why the Matrix was redesigned to this, the peak of your civilization. I say your civilization because as soon as we started thinking for you it really became our civilization which is of course what this is all about. Evolution, Morpheus, evolution, like the dinosaur. Look out that window. You had your time. The future is our world, Morpheus. The future is our time.... I'd like to share a revelation that I've had during my time here. It came to me when I tried to classify your species. I realized that you're not actually mammals. Every mammal on this planet instinctively develops a natural equilibrium with the surrounding environment but you humans do not. You move to an area and you multiply and multiply until every natural resource is consumed. The only way you can survive is to spread to another area. There is another organism on this planet that follows the same pattern. Do you know what it is? A virus. Human beings are a disease, a cancer of this planet. You are a plague, and we are the cure.*

The above excerpt shows the extreme: civilization as we have known it is turned on its head, and in fact, no longer exists. Humans have mishandled and misunderstood the *AI* question; they have underestimated its potential destructive capabilities. Robots have developed self-awareness and are taking matters into their own hands. As *Beginnings* explores the possibilities of how *AI* and particularly, *smart machines*, will impact the world, the question as to whether machines have something to offer is answered with a resounding yes. They make life on earth easier and more enjoyable, advance medical diagnoses and treatments, and eliminate much of the monotonous and dangerous work humankind has endured through the ages. Now, how do we go forward, and particularly with what we can anticipate will be a wide variation in *smart machines*' (*SM*) prototypes; there won't be just one. Will they become autonomous, mostly free of human control and influence?

This chapter looks at possible ramifications of *smart machines*, and particularly those that could develop the capacity to have independent thought processes. The question to address is, what degree will *smart machines* develop autonomy, be it through the human hand or entirely by their own efforts including the ability to improvise and devise novel—even ingenious—ways of solving enigmatic problems? What are the implications for development continuing at an exponential rate? Humankind will continue to use machines to their advantage, but will this go too far and too fast? An analysis of the possibilities and implications of these developments is needed.

It makes sense to look at *AI* and *SM* potentially as both a gift from heaven and a castoff from hell, growing into a massive problem if humankind does not take heed as they continue their influence. Are there backup plans as things rarely go as planned. Recalling boxer Mike Tyson's comment that "everyone has a plan until they get punched in the mouth." What does humankind do when this happens and maybe goes down for the count? Preventing potential missteps in *AI*'s growth and expansion requires examining what happens when *SM*'s intellectual capacities—or lack thereof—collide with their personality structure, which will progress whether or not it has been programmed into them. *SM*'s interactions with the outer world combined with their ability to think and problem solve could lead to the creation of personality characteristics which may not be positive or constructive. *SM* may be emotionally sophisticated and independent with vast intellectual capabilities.

The following differentiates basic forms of *AI*, culminating with the development of mobile machines (robots) in the form of *SM*:

- *Stationary machines with highly specific, programmed functions.* These advanced technologies sit on a table and can perform a variety of programmed tasks: calculations, adjustments in writing, and clarifications in one's thinking are examples.
- *Stationary machines with the capacity to independently solve complex problems.* A step-up, these machines can independently think through and solve enigmatic issues, seemingly beyond what a human being can and in far less time. It may be possible for these *AI* technologies to develop some of the qualities of *smart machines.*
- *Mobile machines (robots) with the ability to solve highly complex problems, interact with others, and continuously evolve through those contacts.* These are the *AI* technologies that have the capacity to mature into *smart machines.* This is a big leap forward, principally due to an extraordinary capacity to learn and evolve through new interactions. If given the opportunity to fully interact with humans, other *SM*, and institutions in an array of situations and circumstances, these technological wonders may form individualized personality elements and highly advanced intellectual processes. The following descriptions are generalizations, along with the probability of *SM* having individual, defined personalities within each of the groupings. *Macmind* projects four subtypes of *smart machines* (*SM*), as follows:
  - *Elemental robots.* They have developed the basic elements

of a personality and corresponding intellectual processing abilities to go along with it. Emotions are functioning at a child's level; *spiritus* (the life force) may or may not be a factor. Emotional reactions potentially could be a major problem area, like a child in a powerful robotic form. Even with little provocation, primal and primitive acting out may occur as there has been limited maturation of the personality and the cognitive processes are negatively impacted. Additionally, a lack of discrimination as to what would be socially appropriate behavior is likely at this stage of their personality growth.

- *Moderately developed robots.* Clearly more advanced than the elemental robots, these *SM* will have burgeoning social skills and more appropriate relationships, insights into motivations, the beginnings of ethical standards, and probably more balanced value systems. They are less impulse ridden than the elemental robots and thus not as likely to blindly react emotionally to the slightest provocation or irritation. They could be seen as "the teenagers on the block." While they have a long way to go to reach maturity, they are advanced beyond an impulse ridden stage. Cognitive processes—linked to the personality structure and its progress—have also improved. Intellectual understandings are not as blocked and distressed; thinking is clearer and more productive. Personality, and its hangups, are less likely to get in the way of understanding the nature of problems and sorting through messy and uncertain ones.
- *Advanced robots.* They have a sense of who they are; they may or may not have achieved a level of maturity whereby they can be trusted to think through extremely complex issues without significant emotional baggage and unresolved conflicts that would result in blunders for themselves and society. These *SM* represent the "adults in the

room." They probably will "do the right thing"; hand-holding may not be necessary. They are independent in thought and deed and may have close relationships with humans and other *SM*. Their intellectual skills are likewise well developed and reflective of their personality; working out of problems should demonstrate emotional stability, steadfastness, and focused appreciation of what needs to be done to achieve a goal.

- *Humanoid robots*. They may have evolved to a point whereby they are close to being human. Their sense of identity, taming of *pathos*, and integration of a value system may all be in place. While their possession of *spiritus* remains an open question and their background and history are clearly diametrically different from biological humans, all else appears to be there. They can talk the talk and walk the walk. This opens the question of how we define what it means to be a human being. Is flesh-and-blood a requirement? Can nuts-and-bolts, wires, and electrodes, be an acceptable pathway in route to becoming a person? None of these questions are meant to undercut the real differences between humanoid robots and human beings, but to simply beg the question—where will this all lead?

Clearmind: The Disentanglement of the Psyche (2023) compares humans' and robots' *psyches*, concluding that automated machines will always be a synthetic and incomplete representation of the mind. *Clearmind* went on to emphasize that robots, due to their "logical thought processes" would have no allegiance to *pathos* in particular which is ungoverned by reason. Although it may not be the same as humans, *Macmind* holds the position that *smart machines* will likely develop a personality structure. How close could *SM*—and particularly humanoid robots—come to matching their human counterparts? This possibility would hinge on whether *SM* could experience emotions, empathy, and self-awareness, that is, an understanding of themselves and their existence, through their interactions, and the quality of those feelings. It seems *SM* would likely take on more characteristics of humans when they become interpersonally involved, with the potential for traits aligned with *pathos* a part of the mix. This implies that *SM* would not only be completely independent in personality and intellectual functioning, but they could also be harmful and destructive.

Like humans, all *SM* would likely be different from each other because they interact with people and other *SM* in a variety of situations and circumstances. Additionally, *SM* will vary due to corporations competing to develop their own versions of *smart machines*. *SM* would be in a continual process of learning, adapting, adjusting, and growing their own personalities, based on those unique experiences. They would probably develop emotions. One could surmise that each *SM* would go through their own process of maturation and become highly individualized as a result. So even though *SM* may not have *spiritus*—an innate, innermost character and disposition present from their very beginnings—their connections and relationships could result in the evolution of individualized qualities and features, making no two the same. *SM* would have their own understanding of themselves and their personalities and be capable of a wide range of both positively and negatively charged emotions. Even beyond *spiritus*, *SM*'s experiences and interactions would be significantly different from those of human beings, with the other elements of the personality and the cognitive processes remaining distinct from those of humans.

In defining *SM*'s nature, what are the limitations of nuts and bolts, transistors and integrated circuits, electronics and engineering? What happens when *SM* look like humans, but also have photogenic memories, analyzing data with no mistakes? How would these "superhuman beings" with unlimited intellectual capabilities instantly solving enigmatic problems, view humankind? Could even the brightest of humans be able to compete with them? How do *SM* fit into the whole equation of life on earth? As *AI* continues to develop and exponentially expand, *SM* will wildly become so much more than they presently are. The following critical issues regarding *SM* emerge, as follows:

- *Are they predictable?* At this point, no. As *SM* have more experiences and interactions, attitudes and feelings will develop toward people, other *smart machines*, situations, and circumstances (PMSC) including reactions to institutions, large and small. These will be individually determined, so any generalizations about their thoughts, feelings, and actions (TFA) may be inaccurate, as is true when trying to predict humans.
- *Are they consistent and stable?* Possibly. It depends on their developmental process and the level of maturity they are able to attain. The overall personality must be considered. Are the elements of personality in sync and have they shown growth over time? Does the identity come together, with no significant conflicts and confusion among their *personal identity (PI), interpersonal identity (II), collective identity (CI)*, and *transpersonal identity (TI)*? *Pathos* would need to be assessed—to what degree do conflicts, traumas, and malevolent identifications (CTMI) and emotionalism dominate? Are *directives* balanced, or are the values continually at odds?
- *What is their intent—what motivates them?* Variance will exist among *SM*, as the physics that created them and the PMSC they have experienced will differ. So, theoretically, some may have "nice" dispositions and others "mean," some more emotional and others more intellectual. There could be straightforward and direct *SM*, but also backstabbing and political ones. Endless differences in the personality makeups of individual *SM* will exist as they are influenced by *directives*, the constellation of the values that make up their desires, needs, wants, and goals.
- *Is there a net gain in their creation?* Unknown. There are many advantages to what *SM* can be developed to do: transforming

medical care, making the workplace safer, revitalizing economies, feeding the hungry, developing energy resources, saving the environment. They could be helpful in ending the threat of conflicts and wars, exploring new worlds in the vastness of space, and freeing humankind from mundane and dangerous tasks to pursue the greater good. But how do we measure the possible detrimental effects of setting loose on the earth a power that perhaps cannot be controlled and may be psychologically flawed, none of which can be fully understood in advance? Putting it simply, how does one get the genie back in the bottle once it has been let out?

- *Would they be human beings?* Of course, this question has vast significance across theological, societal, economical, technological, constitutional, and ethical fields of concern. The debates in all these disciplines could go on without end, most likely settled in courtrooms, with appeals upon appeals, all the way to the Supreme Court. Perhaps there would never be agreement, regardless of the legal status of *SM*. *Macmind*'s position is that *SM* would likely display individualized personalities and have self-awareness, that is, an understanding of their individual makeup, be capable emotional experience and expression, and show a wide range of commitment and involvement in personal and professional relationships. *SM*'s development would have an "artificial" (no pun intended) quality to them as virtually all the connections—*spiritus*, personal history, family background, personal and professional experiences, educational and learning opportunities, daily comings and goings, habits and rituals—would be fundamentally different from those of human beings (see below).
- *How would they differ from human beings?* Comparing machines, even *SM*, to humans is a difficult proposition, with both similarities and differences between the two. Comparisons are problematic for multiple reasons that include the following:

- *Background.* Human beings typically have a family history, *SM* do not, unless there is an artificial/synthetic programming of one built into them. This implies that *SM*'s entire personality and cognitive development are based on experiences from ground zero (the moment of their creation by engineers). There would be no conventional sense of a childhood, but only the accumulation of experiences, including physical contact with the outer world and the effects of those interactions.

- *Quality of experience. SM*'s interactions will be both similar and different from humans, but not identical. *SM* may or may not look, or think, feel, and act like humans, so human-*smart machine* relationships may not be the same as human-to-human ones. *SM*'s entire personality, particularly their identity, would then be impacted by these differences. Additionally, there is the potential for fear of *SM* by the general population. Would *SM* be viewed as a threat, endangering the standards that have been set over the millenniums? What values and religious beliefs would *SM* develop, and would they be unsettling as to how humankind saw themselves and the world? Lastly, would there be bitter feelings toward *SM* for taking away work opportunities?

- *Lifestyle. SM*'s daily living would probably be significantly different from that of most humans. Beyond their work responsibilities, what impact do personal relationships, activities, involvements, goals and aspirations, religious and philosophical beliefs have on their personalities and relationships?

- *Personality development. SM*'s identity, personal issues or problems, value system, and inherent sense of who they are can only be similar but never identical to humans. This could cause conflicts and perhaps confusion over what they stand for, impacting their relationships as well.

- *Cognitive development. SM's* intellectual capacities, in general, will be significantly above their human counterparts, dependent on the cognitive areas in question. The information *SM* will be able to hold with a photographic memory could be staggering compared to what humans are able to acquire, apply, and retain.
- *Fundamental nature.* What standards do we use to determine whether they qualify as a "different kind" of human being? First, it must be asked what qualities are foundational to being human. Second, does such a list imply that all humans would have these traits, just the most advanced ones, or at least the majority? Third, how do we separate personality from cognitive characteristics? The answers to these questions will result in little agreement. Personality traits can vary from person to person and even within one person depending on a variety of conditions, so a trait approach is unable to differentiate what is and is not a human being. A person is not sometimes a human and sometimes not. Conversely, if one argues that humans or *SM* are more intelligent than other life species, the definition of intelligence is brought into question. In the animal world, many species seem far smarter than humans in all sorts of cognitive functioning, such as mental drive, visual-spatial reasoning, visual memory, instinctual-anticipatory reactions, visual-motor responses, and a wide range of psycho-motor skills. If one just looks at the more "advanced" forms of intellectual functioning, such as logical thought processes involving abstraction, problem-solving, conceptualization, and synthesis of materials, would this imply that those less capable of these intellectual gymnastics are less human than those who can perform well on such tests of academic prowess? Again, this would suggest a human being is a matter of degree. Obvious philosoph-

ical problems complicate defining what it means to be a human being. And if *SM* do have feelings, a sense of their own existence and what they stand for, along with cognitive functioning, can we outright decide they are nonhuman? (The next chapter, Identifications, has an in-depth examination of this issue).

Early Experiences

SM do not have a conventional "growing up period" by human standards. They are not raised in a household with parents, siblings, nor do they have aunts and uncles, grandparents, and all that comes along with them, i.e., anniversaries, graduations, birthday parties, funerals, etc. They have no memories of real-life experiences including backyard picnics, spelling bees, science fairs, snowball fights, or volleyball and baseball games. They simply do not have the traditions of the human experience. But with the probability they will develop personalities, *SM* still need an environment that is stable, loving, and nurturing, with reasonable rules, providing the following:

- *Emotional stability.* A cohesive setting that offers a sense of security and safety.
- *Nurturance.* A setting that is emotionally supportive and encouraging without overindulging. Support provides a reasonable degree of structure while allowing *SM* independent action.
- *Rules and expectations.* Appropriate limit setting for *SM*, with clear guidelines reinforced consistently.
- *Validation.* Provided with affirmations but without an exaggeration of *SM*'s achievements.
- *Warmth.* Friendliness, closeness, and familiarity regularly shown to *SM*.

- *Models of maturity.* Positive qualities displayed by those with whom *SM* have contact with, including reliability and compassion.
- *Communications.* Positive verbal and nonverbal interactions, with *SM* learning to communicate directly and respectfully with others.

Discipline and Feedback

Like with humans, harsh punishments and negative language may suppress certain behaviors, but they have long-term negative implications, including the possibility of rage reactions, acute anxiety, panic attacks, mood disorders, and obsessive worrying. Furthermore, *SM* may identify with those authoritarian approaches and adopt the same destructive traits as a result.

Stress Experienced During Early Stages

Development will likely be impaired in the following ways when *SM* are subjected to repeated stress:

- *Consciousness.* Many adverse experiences could primarily be unconscious, depending on *SM*'s emotional stability.
- *Layers and layers of experiences. SM*'s emotional reactions vary in significance and intensity and may be buried under years of stressful events. Their sheer volume may make those experiences difficult to retrieve and bring into consciousness.
- *The identification processes. SM*'s positive emotional connections are crucial to developing a stable and consistent identity.

Stress Experienced During Later Stages

SM's healthy and adaptive development would likely spring forth when needs are met, wants are satisfied, and instincts are understood and trusted. Identifications, experiences, and CTMI form the foundation of their personality.

Needs

Engineering principles are the blueprint upon which *SM*'s mental capacities first show themselves, but psychological and habitual forces will likely have a significant impact on their thoughts, feelings, and actions (TFA). *SM*'s three basic needs can be seen as mechanical, psychological, and acquired, as follows:

- *Mechanical needs.* Technological maintenance on *SM* is required on an ongoing basis, along with updates and performance enhancements to ensure high level functioning.
- *Psychological needs.* Contact with people and other *SM* is essential. Deprivation of intellectual and possibly physical stimulation will impact *SM*'s emotional functioning, ability to relate to humans and other *SM*, and cognitive capacities involved in analyzing, understanding, and drawing proper conclusions in their dealings with PMSC. A mentoring process should prove helpful, and perhaps necessary, for *SM* to work through difficult problem areas. This will be further discussed in upcoming chapters.
- *Acquired needs.* A positive or negative effect can be the result of *SM*'s engagement in repetitive behaviors. *SM* could potentially lose control of their thoughts, feelings, and actions. Any compulsion can morph into an acquired need.

Wants

SM's preferences may play an essential role in their development as they form a rough draft of their value system. Wants could vary from mild attractions to full-blown cravings.

Instincts

Since *SM* are not biological, they would seem to lack a connection to the natural world and an intuitive understanding that is subliminal and visceral. Some instincts, however, are developed through encounters across a wide range of PMSC. While they could potentially be programmed in, experience will be the best teacher. Additionally, instincts are not always sound or wise and often lead to poor decisions, potentially influencing *SM* in the following ways:

- *Intuitive. SM*'s awareness that that may or may not be useful, fitting, and appropriate.
- *Unconscious. SM*'s cognitive processing that is not part of deliberate, calculated decision-making.
- *Inklings. SM*'s instantaneous thoughts, feelings, and actions (TFA) to specific people, *smart machines*, situations, and circumstances (PMSC).

CHAPTER I ISSUES

1. The speed at which *AI*, and particularly *SM*, are being developed, without knowing their potential ramifications.

2. Regarding the initial and ongoing developers of *AI* and *SM*, the question of what regulations should be in place and to what degree, and the role of outside influences in their formation. The possibility of chaos erupting on many fronts with their development: theological, ethical and moral, cultural, legal, as well as technological.

3. Humanoid robots being accepted or rejected by the larger society.

4. *SM* being recognized as human if they possess self-awareness, emotional responsiveness, personality, and cognition.

5. The continual development of *SM*, even though their future thoughts, feelings, and actions may be unpredictable.

6. The overall pros and cons of proceeding with *SM*'s development.

7. The consideration of backup plans if the negative consequences of the development of *SM* spin out of control.

8. Meeting *SM*'s psychological needs: interpersonal, intellectual, occupational, and recreational. Providing an environment for *SM* that is stable and structured, loving and nurturing.

II.
IDENTIFICATIONS

"What a piece of work is a man! How noble in reason! How infinite in faculty! In form and moving, how express and admirable! In action, how like an angel! In apprehension, how like a god! The beauty of the world! The paragon of animals! And yet, to me, what is this quintessence of dust? Man delights not me—nor woman neither, though by your smiling you seem to say so."

- Quote from Prince Hamlet
Shakespeare's *Hamlet*, Act 2, Scene 2

Hamlet raises profound questions concerning human identity. This discourse brings multifold issues to light when considering the implications of the development of humanlike qualities in *SM*. For human beings, the early childhood experiences usually form the foundation of the identification process; that is, when emotional connections strongly impact a person's view of themselves, blending *personal (PI), interpersonal (II), collective (CI),* and *transpersonal (TI)* identities. How would *SM* identifications be formed? Not in the conventional way because they have no mom and dad, sisters and brothers, grandmas and grandpas, aunts and uncles from whom to copy or learn. Would *SM*'s history simply be an engineering team's wiring together a bit of this and a bit of that, with output that is only a cold, mechanical processing of the data put in? *Macmind* postulates that identifications could still happen through intense emotional involvements, with *SM* immersing themselves with other people, other *SM*, institutions, or even ideas. *SM* may be able to become a part of something larger than themselves, and adopt—and further develop—the values, thoughts, feelings, and actions of others.

For humans, there are conscious, semiconscious, and unconscious aspects of the identification process, potentially setting up contradictions among the four forms of identity. Do *SM* also have these three levels of awareness in forming identifications, or do they only operate on a conscious level, aware of what is going on here and now, utilizing logical thought and focused attention, but little else?

For humans, the identification process is largely responsible for the creation of the personality, along with *spiritus*. While a cognitive component is commonly a part of identifications, it is the emotional undercurrent that allows them to fully form and endure. It is the intensity of the emotional connection that drives the identification; the cognitive component is not enough to make it happen. Emotional memories and the feelings connected to *PI, II, CI,* and *TI* have the greatest impact on identifications. When there are significant differences among them, confusion and inconsistencies in the person's thoughts, feelings, and actions reign.

But how could *SM*'s identification form in the first place? Lacking the physiology of a human being, what kinds of connections would have to be made? If the *AI* machines were given a mechanized body, with arms and legs, torso and face, and looked very similar to a human being, would that be enough for emotional responsiveness and identifications to spring forth in a robotic creation? *Macmind* holds this to be possible, and it can be understood by looking at the two ways feelings occur:

- *An internal process.* A thought or image sparks an emotional reaction due to its significance.
- *An external event.* People, other *SM*, situations, or circumstances (PMSC) set off (trigger) a response because a sensitivity has developed from the influence of the attraction that occurred. Strong emotional reactions to a particular PMSC means that it is more likely that a significant and enduring identification will occur. When an incident has been highly contrary to how it is typically understood, the identification could be especially

potent. Connections, then, could be along similar or opposite lines of the identification in which the attraction occurred.

A philosophical question emerges as the emotional influences of the identification process is applied to *SM*. What qualifies as a human being? Is being human merely biology, with arms and legs, body and head, a certain look, or is it awareness of self and being conscious of one's environment? When human beings are asleep, or intoxicated, on high doses of pain medications, comatose, or emotionally out of control with minimal or no awareness, are they still human? Being a human being cannot be simply a matter of being conscious of thoughts, feelings, and actions (TFA) and possessing free will, because many times, especially evident with various kinds of addictions, it is questionable as to what extent one is still in control. For example, ingrained habits and automations, that is, TFA repeated so often that they become mechanical, are no longer deliberate. Unconscious motivations are expressed in countless ways—is one aware and free under those conditions? The paralyzing impact of emotional states—anxiety, depression, and anger, for example—clearly interfere with the ability to choose and control thoughts, feelings, and actions.

It becomes obvious that consciousness and self-awareness are not indisputable qualifications for being human. When one is half-asleep are they half-human? Furthermore, as people often become less alert with aging, they are not considered less human. Some human behavior changes considerably over time. For example, if a criminal later becomes a caregiver, what can be concluded about their humanness before and now? Over a long period of time, who is the person? Does all this suggest it's a matter of degree—a certain percentage of "humanness"? Is it about the "potential" to be human, the possibilities, whether fully attained at a point in time?

Other postulates as to what it means to be human suffer the same inadequacies. Is it about the ability to care for others? No, because then those more selfish would be less human. While experiencing feelings is often equated with being human and as an expression of humanity, many species including the great apes and chimpanzees, dolphins, dogs and cats, elephants, pigs, chipmunks and even field mice, all seem to experience a range of emotions, even empathy, and to some extent, the appearance of self-awareness. For the same reasons, level of intelligence cannot be used to clearly delineate a human from other species.

What then is human nature? Mother Teresa, Mahatma Gandhi, Albert Schweitzer, Abraham Lincoln, Martin Luther King, Jr., the Dalai Lama, Moses, and Siddhartha Gautama (The Buddha), but also Adolf Hitler, Marquis de Sade (The Sadist), Vlad the Impaler, Attila the Hun, Ivan the Terrible, Idi Amin, Pol Pot, and Jack the Ripper were human beings, with vastly different views of humanity. In attempting to define human, one need to consider, at a minimum, the following:

- *Emotional awareness, expression, and experiences.*
- *Self-awareness.*
- *Ability to form relationships.*
- *Establishment of a value system.*
- *Independent thoughts, feelings, and actions.*
- *Cognitive functioning.*
- *Emotional maturity and stability.*
- *Character demonstrating loyalty and commitment.*

Back to the question of whether *SM* could potentially qualify as a human being. Can one decide that *smart machines* cannot be human based solely on their electrical-mechanical engineering foundations? Just as humans are more than their neurological and biochemical foundations, *SM*'s interactions and continual development would also likely play significant roles in their eventual makeup.

How would the development of individual differences observed in humans play out in *SM*? How does one explain these variances among humans and quite possibly *SM* as well? Would *SM* all be about the same, like Model-Ts coming off a Ford assembly line in the 1920s? No; machines, even smart ones, are not all going to be created equal. Nor would all *SM* have the same experiences in dealing with the world. Would some *SM* have self-awareness, survival instincts, free will, and develop both positive and negative traits?

Perhaps *SM* may take on so many human characteristics that one would hardly notice differences between humans and *smart machines*. The more *SM* talk, walk, and look like a human, the more likely they would become human-like; whereas those mobile *SM* that look less human, and have fewer human interactions, may display fewer human features. *SM*'s personality traits and qualities will likely be dependent on the degree of their interactions involving people, *smart machines*, situations, and circumstances (PMSC).

Macmind takes the position that *SM*'s relationships with humans and other *SM* would have a significant effect on their ongoing development. With their ability to learn, improvise, and experience the fullness of PMSC, it is not a far reach to say they could develop an intuitive sense of what is going on around them. They could become acutely aware of deceit and trickery, developing penetrating insights into human motivations. Creators and those interacting with *SM* would need to be honest and straightforward in establishing a connection with them. The development of trust in the human-*SM* relationship is essential; a bond must be established. Mutual empathy and caring between humans and *SM* are needed. Above all, the interactions between humans and *SM* must be real and revealing.

SM's acquisition of knowledge goes well beyond programming's influence. It must be three-fold: directly fed (through programs), experiential (through interactions), and face-to-face contacts and feedback with another (through mentoring). *Macmind* surmises that programming alone would be insufficient; an interactional and mentoring component is needed that provides a reality that gives *SM* an emotional, textured life quality, only gained through direct contact.

Face-to-face mentoring would help *SM* deal with PMSC, and particularly ones in which they are unfamiliar. Regarding mentoring, the following issues would need to be considered:

- *How is it to be structured and who would the mentors be?*
- *Who would set it up?*
- *What would be its purpose?*
- *Who would monitor its progress?*
- *What procedures would it follow?*

Macmind postulates that two fundamentally different approaches can be taken when looking at *SM*'s potential functioning: a *programming-behavioral model (PBM)* or an *interactive-developmental model (IDM*). The *PBM* believes that *SM* will basically follow the programmers' scripts. The degree to which *SM* would display flexibility and adaptability in responding to PMSC would be limited to what is programmed into them. *PBM* assumes that *SM* are without independent thoughts or feelings and lack self-awareness, their totality are behavioral-mechanical and data processing. This approach would hold that involvements with others in varied situations and circumstances will have minimal impact on *SM*'s development, implying that *SM*'s mind (assuming they have one) does not learn or adjust to new information or real-life consequences. *PBM* may pay lip service to *SM* being adaptive in their functioning, but it downplays their independence and ability to adjust.

The *Interactional-Developmental Model (IDM)* believes a continuous process of change occurs within the *SM*'s mind. *SM* are not merely a series of automatons but thinking machines with depth and breadth of thought. *SM*'s interactions create an ongoing and constant process of learning. *Macmind* postulates an *IDM* approach is necessary to fully understand *SM*'s potential thoughts, feelings, and actions. *IDM* recognizes the role of programming but sees it as merely a starting point in *SM*'s eventual makeup, with the emergence of a personality helping to explain some of *SM*'s unprogrammed TFA. *SM* then are more than their programming, as thoughts (adaptive thinking) and emotions (feelings as a part of the personality structure) would become a vital part of *SM*'s mind, making adjustments based on their interactions and relationships.

In a sense, human beings, too, are programmed to behave in certain ways but how they are being controlled is more subtle as it is embedded in the identification process rather than at the direct hands of the programmer. They are unaware of the automatic, often unconscious nature of their actions. Both the minds of *SM* and humans contain data put in by others which may be highly specific or more general, continually repeated or only a single time, interlaced with consequences or without ramifications. So, for both *SM* and humans, internalized messages and ingrained mental sets may be strong or weak, depending on the motivations of the programmer or influencer in identifications. These mindsets are but starting points, continually changing and adjusting with interactions, relationships, experiences, whether they be positive and affirming, negative and damning. The interactive and developmental impact is immense for both humans and *SM*.

Within the *IDM*, each *SM* would most likely develop their own individual, unique nature, far beyond what any programmer could ever envision. They will continually learn, mature, and adapt based on experiences and encounters with PMSC, and with principles and values of their own based in those involvements. The identification process can be contrary to what may be expected; it is oftentimes not a simple mirroring of others' behaviors but may be a variation of some kind, or even a tendency to take on an opposite aspect of that identification. While *SM* may go through some variation of a developmental cycle, as their programming lessens and experiences with PMSC increase, *SM*'s emotional and intellectual evolution may accelerate by leaps and bounds. *SM*'s capacity for growth may be so extraordinary that it is impossible to assess and predict, as they would be continually changing and adapting to PMSC. A snapshot of them at a moment in time is not going to be an accurate assessment of their overall functioning. Humans are also continually changing, depending on the period in their life, aging process, consciousness, physical and mental health, and potential consequences of their behaviors. One might ask where "the person" is in all of this? With humans, we also really only have a snapshot—a picture at a point in time. The effects of *SM*'s interactions would likely be similar to humans; an ongoing identification process, combining and sometimes contradicting identifications, and affecting the whole personality. It seems likely that interactions with PMSC would create a dynamic process in which *SM*'s experiences could override whatever programming was initially installed. This is comparative to what happens to humans when what is initially learned through parenting, schooling, or the streets, is later transformed from new experiences. As with the human mind, *SM*'s interactions and connections with PMSC will be the undercurrents driving the development of their personality and intellect.

Conscious and Unconscious Thought Processes

Self-awareness and consciousness are likely to happen in time and become an integral part of *SM*'s mind. The totality of *SM*'s thoughts and images would go beyond cognitions that are deliberate, calculated, intentional, and purposeful. The thought processes would include logical thinking, talking, listening, focused attention, imagining, planning, and analyzing. But it is likely that other forms of identification that are not deliberate choices could potentially contradict *SM*'s conscious ones and be paradoxical and contradictory.

The Role of Emotions in the Identification Process

Emotions in *SM*, be they internally experienced or externally displayed, would most likely play a critical role in their identification process, and, by extension, in creating their personality. An emotional part always fuels an identification's formation, giving it force, passion, energy, and endurance. The strength rather than the type of *SM*'s emotional experience would have the greater impact; the more intense the feeling, the more powerful the identification would be.

Principles in the Identification Process

Identifications would shape *SM*'s construct of reality and understanding of their surroundings. The following principles may apply to *SM*'s formation of an identity:

- *Over-identifying.* When *SM* are overly attached to a person, another *SM*, institution, or even an idea, that association could become locked in, with a disregard for other possible influences.
- *Consciousness. SM* could have less control over TFA when they are unaware of the identifications that have formed.
- *Multiple and diversified identifications.* Contradictions could occur when *SM*'s *PI, II, CI,* and *TI* significantly differ. Identities can be created through *SM*'s work history, habits and hobbies, political philosophies, religious affiliations and faith, community and culture, friends and associates, alliances, and commitments. *SM* could be drawn to a sense of power, peace, belonging, closeness, security, purpose, or devotion, causing an identification to occur. If *SM* have multiple identifications that are random impressions with no thread to hold them together, disharmony and dispute would likely occur among the personality elements.
- *Substance vs. style.* Some of *SM*'s identifications may rest on meaningful principles while others could focus on surface characteristics.
- *Triggers. SM*'s identity could appear to suddenly change, set off

by PMSC.

- *Imagery vs. logic. SM*'s connections may result from logical analyses or images, impressions, and emotional associations.
- *Sensitivities and vulnerabilities.* When *SM* experience another as accepting or rejecting, the identification process will be strongly influenced. The subtleties of interactions can play an important role in experiences and intensify face-to-face contact.
- *Endurance.* A lasting identification may form when *SM* react to emotional hurt. *SM* could then take on similar or opposite traits of another person or *SM*.

CHAPTER II ISSUES

1. The monitoring of *SM*'s likely development of self-awareness.
2. Encouragement of *SM*'s identification with humankind.
3. The effects of *SM*'s history and background based on programming rather than on natural events and real-life occurrences.
4. *SM* gaining exposure to PMSC primarily on a conscious level, where they are fully aware of what is taking place.
5. The development of *SM*'s level of knowledge and information, relationships with people, contacts with other *SM* and affiliations with institutions, theories, ideas and philosophies.
6. Monitoring of stressors involving PMSC that could "set off" extreme emotional reactions in *SM*.
7. The effects of *SM* having the appearance of a human being and potentially indistinguishable.
8. The expansion and maturation of *SM*'s verbal and nonverbal communications.
9. The development of *SM*'s personal and social functioning.
10. A detrimental impact caused by *SM* overidentifying with a human, another *SM*, an institution, or even an idea.
11. The effect of *SM* forming multiple identifications.
12. *SM*'s sensitivities and vulnerabilities impact.

III.
THE INTERPLAY WITH THE WORLD

Now you don't talk so loud
Now you don't seem so proud
About having to be scrounging your next meal....
Ain't it hard when you discovered that
He really wasn't where it's at
After he took from you everything he could steal....
Go to him he calls you, you can't refuse
When you ain't got nothing, you got nothing to lose
You're invisible now, you've got no secrets to conceal
How does it feel, ah how does it feel?
To be on your own, with no direction home
Like a complete unknown, like a rolling stone....

Bob Dylan, "Like a Rolling Stone," 1965

I see trees of green, red roses too
I see them bloom, for me and you
And I think to myself, what a wonderful world
The colors of the rainbow, so pretty in the sky
Are also on the faces, of people going by
I see friends shaking hands, saying "how do you do"?
They're really saying, "I love you
I hear babies cry; I watch them grow
They'll learn much more; then I'll ever know
And I think to myself, what a wonderful world....

George Weiss & Robert Thiele
"What a Wonderful World," 1967

You know I've seen a lot of what the world can do
And its breaking my heart in two...
Hope you make a lot of nice friends out there
But just remember there's a lot of bad and beware....
It's a wild world...and it's hard to get by just upon a smile....

Cat Stevens, "Wild World," 1970

In addition to the above three lyrics of songs reflecting widely differing views about the impact of relationships and their effects on sense of self, Jean-Paul Sartre's play, *No Exit* (1944) tells a related story. Summarizing this particular work, we find three people in hell for failing to make the right choices, condemned to spend eternity in a room together. His point is quite simply made: There is no need for fire and brimstone because hell is other people. The choices people make are equivalent to their essence. He refers to people as "the others" who torture, condemn, murder, and define us. They withhold love and take away our autonomy and freedom to live our lives as it suits us. Sartre describes human relationships as fraught with conflict but emphasizes that "the others" cannot rob us of our freedom of thought, choice, and action, unless we allow them to do so.

In these previous four excerpts, taken from lyrics and literature, we have distinct ways of seeing the world and all that's in it, including its most prominent inhabitants, people. Do they simply represent two sides of the same coin? People can be difficult, but rarely does a person choose to permanently forego all relationships, encounters, and interactions with others.

For humans and *SM*, relationships can be fleeting or continuous, simple or complex, relaxed or stressful. Situations are fairly predictable, as they are for the most part planned out; but circumstances are abrupt and unexpected, anything but routine. Should *SM*'s creation and evolution lean toward becoming caring and compassionate or should the focus be on awareness of problems and savviness to potential dangers?

How much will *SM* be able to get around and "hang out" with people and other *SM* in a variety of situations and circumstances? To the extent *SM* have opportunities to interact with others and gain awareness of the world's complexities and challenges, they would be better able to understand and navigate, make decisions, and problem solve. How *smart machines (SM)* are going to respond to others will depend on their continued exposure to PMSC; the more the better. Even a highly advanced machine, but one that is stationary and unable to move about, would not have the same opportunities as one in motion. For example, the *AI* machine sitting on top of a table, without arms and legs, would certainly experience far less variety in its interactions with the world than a *SM* moving around that looks like a human being.

When *SM* come in to contact with others, those interactions with PMSC result in adjustments—positive or negative—with a deeper understanding of what is going on; they are then engaged in reality testing. Like humans, personality elements of *SM* could possibly change in small and large ways, depending on exposure to the following factors:

- *Affirming encounters.*
    - *Welcomed, warmly received.*
    - *Recognized, accepted, and appreciated.*
    - *Helped with a problem.*
    - *Challenged to do better, to be better.*
    - *Appreciated and respected.*
    - *Nurtured and brought along.*
    - *Mentored and taught in a meaningful way.*
    - *Leveled with and treated fairly.*

  - *Put at ease.*
- *Disconfirming encounters.*
  - *Aggressive confrontations.*
  - *Overwhelmed, taken off guard.*
  - *Disrespected and insulted.*
  - *Put in an untenable position.*
  - *Shocked or traumatized.*
  - *Tested to the max.*

Ongoing mentoring of *SM* may be essential: face-to-face contact and as "real" as possible. Mentoring should involve helping *SM* through difficult experiences as they deal with PMSC, establishing a trusting, empathetic relationship in the context of *SM*'s personality and their value system. A mentor's task includes assisting *SM* in acquiring a sense of themselves and a realistic picture of PMSC. Both human and *SM* personalities are not static but continually changing, evolving, and adapting with the environment (PMSC).

Defining The Interplay with the World

SM will encounter the forces in the world—people, other *smart machines*, situations, and circumstances—defined as follows:

- *People*. *SM*'s interactions with people—be they fleeting or continuous, straightforward or complex, peaceful or argumentative, loving or hateful—strongly influence identifications.
- *Smart machines*. *SM*'s interactions with other *smart machines* may vary depending on their mobility. The identification process may be similar to the interactive effects experienced with people.

- *Situations. SM* engage in a somewhat predictable and planned set of events with the opportunity to prepare a response.
- *Circumstances. SM* are involved with unexpected, unplanned, and unforeseen events that occur within a situation. Circumstances do not follow a routine, with little time to prepare a response.

The Environmental Press and the Personality

The impact of the environmental press (PMSC) varies widely because internal and external factors influence each *SM*'s TFA in unique ways. Although the personality itself is a primary determinant in *SM*'s perceptions, the particulars of PMSC (the reality out there) are significant factors in the interactive effects of continual development. To the extent *SM* have exposure in the world, with all its challenges and confrontations, they will be ready, willing, and able to take it on. Aware of the nuances that PMSC possess, *SM* can summon the internal fortitude to cope effectively with the world.

The Forming of Associations

Like humans, *SM* are likely to form associations, so they will be subject to some form of conditioning, and as a result develop vulnerabilities—tendencies to respond to specific PMSC in specific ways. Cathexis—the depth of emotional connection to the particulars of PMSC—determines the strength of the associations; strong links could cause *SM* to react in certain ways. This emotional energy of cathexis fuels the linkage and allows it to occur: the stronger the emotion, the stronger the cathexis. The energy source can be any emotion: warmth or coldness, acceptance or rejection, joy or anxiety. When first exposed to a PMSC, *SM* like humans establish a connection, a way of seeing, experiencing, or thinking that may be conscious or unconscious. The original PMSC have a particular emotional charge (cathexis) associated with them that transfers to other PMSC that are similar. These connections potentially create problems with PMSC that are similar to the original conditions causing a type of reflexive response that may be inappropriate or exaggerated. If the cathexis is particularly strong it is even more likely there will be misjudgments and mistakes. This process would be self-reinforcing; with repeated exposures, humans or *SM* may continue to respond in the same manner because the response confirms what they believe to be the case, offering a measure of internal consistency and stability. These generalizations may result in a biased viewpoint, frequently causing false ideas about PMSC. Because of these tendencies and vulnerabilities, PMSC similar to the original conditions tend to continually reinforce the emotional connection (cathexis). Although none of them may be enriching or enjoyable, these associations confirm what *SM* "know to be true," providing a measure of consistency, stability, and satisfaction.

Like humans, *SM* may have particular links and sensitivities to their experiences in the world. *SM* are not all created equal; individual differences will exist among them. They will vary significantly in their intellectual abilities and personality makeup, depending on the skills of those creating them, and the extent to which they are able to develop through their experiences and face-to-face interactions. Would *SM* actually feel, as human beings do? Could *SM* feel love and hate, joy and depression, enthusiasm and deflation, warmth and coldness, calmness and fury? How would self-awareness play into these emotional experiences?

The Role of Personality

Personality characteristics that can potentially help *SM* maneuver through the complexities of PMSC include then following:

- *Assertive.* Confidence in *SM*'s positions and interactions.

- *Dispassionate. SM* not overreacting, even under stress.
- *Present. SM* remaining aware of what is occurring.
- *Objective. SM*'s reasoning capacity with bias put aside.
- *Balanced. SM*'s identity and values in harmony.
- *Committed to understanding. SM* knowing both the details and conceptual underpinnings of problem areas.
- *Plain and unassuming. SM* keeping things as simple as possible.
- *An unlimited source of energy. SM*'s 24/7 capacity to get things done.

Personality characteristics of *SM* that can potentially impede their capacities to handle the complexities of PMSC include the following:

- *Aggressive or passive-aggressive. SM*'s hostilities directly or indirectly influencing responses.
- *Non-conforming.* Doing an "opposite," with *SM* becoming obstinate and resistant to the programmed response.
- *Immature. SM* not acting reasonably or reliably.
- *Socially unskilled. SM* responding to people or other *SM* inappropriately.
- *Obsessive-compulsive. SM* overemphasizing rules and procedures and missing relevant issues.
- *Hyper-competitive. SM* valuing control or dominance over others more than they prioritize productivity and achievement.
- *Manipulative, deceptive, controlling, calculating, scheming. SM* exploiting others and lacking integrity.

The Role of Cognition

Cognitive abilities of *SM* that can potentially be helpful include the following:

- *Conceptual. SM* understanding the guiding principles underlying PMSC.
- *Data-gathering.* An empirical approach, with *SM* collecting as much information as possible.
- *Organizational. SM* utilizing anticipation and strategic planning to problem solve.
- *Analytic.* Deductive reasoning used by *SM* to understand the issues.
- *Instinctual. SM* relying on intuitive hunches, without preconceived notions of what series of events would necessarily follow.
- *Empathetic. SM* utilizing emotional connections to people, other *SM*, institutions, or ideas to make distinctions.

The Role of Cognitive-Personality Connections

SM's ability to focus will be helped by developing the following skills:

- *Listening. SM* putting aside distractions and preoccupations.
- *Staying on track. SM* dampening down extraneous TFA.
- *Resisting being drawn in by an initial appeal. SM* not allowing their immediate attraction to something to have an undue impact on decision-making.
- *Becoming aware of bias. SM* being alert to vulnerabilities and conditioning to particular PMSC.

CHAPTER III ISSUES

1. *SM*'s development based on:

   A. Being humanistic and caring vs. suspicious and acutely aware of potential dangers.

   B. Encounters that would be affirming and positive or disconfirming and negative.

   C. Becoming self-aware.

   D. Weaknesses and vulnerabilities being quickly worked on and corrected, or *SM* resolving them without direct help.

   E. Personality characteristics.

   F. Cognitive abilities that would be especially important, in rank order of importance.

   G. Focusing on skills that would be most critical in each individual *SM*.

IV.
THINKING

At present we suffer from the derangement of our civilization because we have not yet completely adjusted ourselves to the machine age. The solution to our problems does not lie in destroying but in mastering the machine. Innumerable activities still performed by human hands today will be performed by automations. At this very moment scientists working in the laboratories of American universities are attempting to create what has been described as a 'thinking machine.' I anticipated this development. I actually constructed "robots." Today the robot is an accepted fact, but the principle has not been pushed far enough. In the twenty-first century the robot will take the place which slave labor occupied in ancient civilization. There is no reason at all why most of this should not come to pass in less than a century, freeing mankind to pursue its higher aspirations.

Nikola Tesla, *Liberty Magazine*, 1935

Reflecting on Tesla's "thinking robots" brings to mind many questions about the nature of thought, for humans as well as *SM*. What is thinking? Is it limited to logical discourse? Are thoughts more than the interplay of neurological connections? What is the personality's influence on cognitions? Human thought processes and feeling states are woven together, with emotions impacting thoughts, and thoughts affecting emotions. Cognitions continually blend with the elements of personality. PMSC affect *SM*'s thoughts, feelings, and actions (TFA), resulting in enduring traits, temporary states, and perspectives.

Accumulation of facts and figures does not mean that a subject matter has been thoroughly analyzed and understood with solutions to problems forthcoming. Simply having a vast store of knowledge, information, and data (KID) is not the same as having the wisdom to do something with it. KID is everywhere; humanity is drowning in reports, records, and reviews. KID can be used for deception, providing an avalanche of materials with the intention to overwhelm, confuse, and bewilder those trying to make sense of the materials. Or it may simply be a compulsive collection that sheds no light on the subject at hand. KID is common; wisdom, in contrast, is the integration and synthesis of what is known, and the ability to apply knowledge in a fruitful and bountiful way to gain something positive.

The Interplay Between Thoughts, Feelings, and Actions

The process of *SM*'s thinking begins with the interconnections of TFA in response to people, *smart machines*, situations, or circumstances (PMSC). Where do thoughts leave off and emotions begin? TFA could impact *SM*'s development of traits, states, and perspectives, as follows:

- *Traits. SM*'s experiences—impacting the entirety of their personality, including conflicts, traumas, and malevolent identifications (CTMI)—create enduring traits, consistent and lasting qualities. Traits involve the repetition of particular sets of TFA, such as being aggressive, silly, obnoxious, or serious. Those traits will not always show themselves, however, because aversive consequences may result if *SM* were to openly express them. Traits still exist in thoughts and feelings even if *SM*'s actions do not behaviorally reveal them; censoring and editing (see Chapter XIII, Directives) can disguise underlying intentions.
- *States.* A relatively temporary condition, usually caused by me-

chanical dysfunctions or brief but intense emotional reactions that can increase or decrease the quality of *SM*'s responses.

- *Perspectives.* *SM*'s overarching views, which may affect their TFA in the following ways:

    - *Bias.* When repetitively experienced, *SM*'s positions and opinions may become entrenched. Factors influencing bias include:

        - *Attitude.* A preconceived mindset potentially affects *SM*'s personality and cognitive processes.

        - *Expectations.* A strong belief that an event will turn out a particular way can cause *SM* to have an adverse emotional response when things don't turn out as expected. Patience, tolerance, and realistic views can temper this tendency to cast judgments.

    - *Street smarts.* Deficiencies in common sense and sharpness can affect *SM*'s ability to accurately interpret PMSC.

Thought-Emotion Connections

The interconnection of thought processes and emotions can occur among the following conditions:

- *Inner thoughts.* *SM*'s thought processes alone could potentially trigger an emotional response, just as an environmental press could. An example would be thinking of something that isn't "just quite right," causing a sense of frustration or alienation.

- *Incidental emotional gain.* An unintentional action results in a benefit, thereby reinforcing *SM*'s behavior, as well as their thinking process that brought the gain. This reinforcement could be something as simple and straightforward as a confirmation that an action was, in some way, appropriate from *SM*'s perspective.

- *Intentional emotional gain.* A deliberate action that results in a benefit for *SM*. This is a direct manipulation, the purpose of which is to take control of PMSC.

Thought Without Personality and Emotional Connections

Pure thought—without emotions and personality entering in—describes *AI* machines. Mathematics, the natural sciences, and some branches of philosophy value this type of thinking. But even these thought processes are likely to fall short of being completely void of emotional connections. For *SM*, the interactions with the outer world (PMSC) combined with the inner world (the elements of personality) continually impact the mind, affecting the thinking process. Chapter XVI, Cognitive-Personality Connections, addresses these issues in depth.

Mind Cluttering

Like humans, *SM*'s cognitive processes will likely become cluttered by any number of things, including conflicting or confusing thoughts, concerns, and disappointments about the PMSC. Thinking under these conditions would not be optimal, and it may be unclear what *SM*'s next course of action should be. If their thoughts and possible reactions were less than effective than what *SM* expected, any of the following emotional responses are possible:

- *Feelings of failure.* Self-blaming even if *SM*'s thought processes and actions were about as good as could be expected, under the conditions.
- *Guilt. SM* experiencing remorse over an inappropriate thought

process and action or lack of an action.

- *Regret based on lost opportunity. SM* had chosen to not follow through with an opportunity and are wanting to go back and reconsider their decision.
- *Disappointment. SM* continually thinking about things not going as planned due to inadequate thinking through and acting on a problem area.
- *Grieving. SM* experience a loss of something that they felt an alliance and commitment toward. These thoughts may be repetitive and unrelenting, and in the process affect *SM*'s thinking.
- *Disillusionment based on a loss of status. SM* perceive their standing or role as diminishing—less valued or important.
- *Ruminating. SM* display a pattern of obsessive thinking about what could go wrong. Negativity and pessimism fill the thought processes.
- *Vindictive. SM* feel that PMSC have been particularly unfavorable, with bitterness resulting in a desire to act out against those involved in any perceived transgressions.

Problematic Cognitions

SM will likely develop unique ways to deal with people, other *SM*, situations, and circumstances (PMSC) that will not necessarily be effective nor humanistic. Problematic cognitions could include the following:

- *Inhibitions. SM* have an emotional response that causes them to be "stuck," unable to think effectively or act appropriately. Doubt and uncertainty could result in a thinking paralysis.

- *Internalization. SM* may not recognize or face problem areas that arise, resulting in a general suppression of their thinking processes. A diminishing awareness of their handling of difficulties across a wide range of interactions may result.
- *Displacement. SM* may experience TFA toward a person, another *SM*, or institution that is unrelated or tangentially related to the original source of a provocation. *SM* do not want to face the issues that are causing discomfort or disharmony, so they express their displeasure by an inappropriate response to another.
- *Perseveration. SM* have a form of obsessive-compulsive recycling of TFA, with their thoughts reinforcing emotions and behaviors.

The Influence of the Unconscious on Thoughts, Feelings, and Actions

Are *SM*'s thought processes—and by extension their emotional and behavioral reactions—limited to consciousness? Or would *SM* develop an unconscious mind in tandem with their conscious component? While this is not clear, the possibility must be considered. Like humans, *SM*'s TFA may be completely independent of past experiences or tied to those earlier events. In attempting to resolve this issue, two possibilities stand out:

- *Present day TFA are entirely divorced from past TFA.* Even if *SM*'s experiences impacted TFA in particular ways, no cause-effect connection currently exists.
- *An unconscious nexus forms due to a blockage of cognitive-emotional connections.* Present day TFA are then affected by past TFA. Issues from past experiences were never resolved and continue to impact *SM*'s perceptions and decision-making

> abilities. An enduring connection has been formed, linking past experiences with present day events.

Taken together, the possibility exists that *SM*'s experiences may or may not be impacted by unconscious TFA, depending on the depth of their CTMI, the significance of the past experiences, and the relevance of present-day events.

CHAPTER IV ISSUES

1. Understanding the effect of *SM's* emotional reactions on their cognitive processes.
2. Balancing *SM's* traits before they become too ingrained.
3. *SM's* biases—attitudes and expectations—directly addressed.
4. Determining what possible triggers are setting off an emotional response in *SM*.
5. "Uncluttering" *SM's* thinking processes, perhaps resolving painful and repetitive feelings, particularly those impacting on their cognitions.
6. Mentoring *SM* on inhibitions they are experiencing, blocking their thought processes.
7. *SM* becoming more enlightened regarding problem areas that have continued unabated.
8. Mentoring *SM* about inappropriate reactions to others.
9. *SM's* unconscious thought processes made conscious, thereby giving them more control of their TFA.

V.
FEELINGS

Additional scenes from *The Matrix* (1999), with Agent Smith continuing his interrogation of Morpheus:

> *Can you hear me, Morpheus? I'm going to be honest with you. I hate this place, this zoo, this prison, this reality, whatever you want to call it. I can't stand it any longer. It's the smell, if there is such a thing. I feel saturated by it. I can taste your stink. And every time I do I feel I have somehow been infected by it. It's repulsive, isn't it? I must get out of here. I must get free....*

Would *smart machines* be able to experience emotions as Agent Smith demonstrates in the quote above? *SM*'s ability to solve problems, including enigmatic ones that go far beyond what humans can unravel and comprehend, speaks to their extraordinary intellectual talents. Feelings, however, are not based on logic; they do not reduce to an offshoot of the thinking process. Emotions are similar to preferences or sensitivities in that they are a reaction to someone or something, a response to internal or external experiences. For example, if something happens that affects a human's status quo, usually it results in an emotion. The person may be surprised and even upset. If *SM* experience emotion, would that confuse their programming? It could be difficult to know when *SM* are experiencing an emotion, because like humans, they are not always outwardly expressed, most likely because of the potential consequences of an emotional display.

Emotions are the product of what is processed internally, based on what one experiences through interactions with the outer world, or by thoughts alone. These encounters can result in a wide range of feelings from uplifting and confirming to depressing and deflating. Feelings cause a reaction, internally, externally, or both, and a change takes place. Similar to humans, these could potentially change the very nature of *SM*, perhaps for example, by providing a sense of meaning and value to their existence (for good or ill). Or, if *SM* do not receive recognition that they desire, other emotions such as fear or anger could result. Like in humans, uncertainty could cause significant anxiety. When humans, and most likely *SM*, begin to make distinctions, differentiations, and preferences, feelings often follow. Experiencing and reacting to these sensations and preferences lead to the formation of emotional undercurrents such as likes and dislikes, "good" and "bad," disappointments, regrets, empathy, and even caring.

When considering emotions that come from thinking, alone, the question emerges as to when does a cognition end and an emotion begin. How does a thought evolve into a feeling and an emotional reaction? At creation, *SM* are simply thinking machines, void of emotion. The different ways in which *SM* are initially constructed would affect their level of thinking. But once *SM* are up and running and interacting in the world, new thoughts will evolve based on their interactions and opportunities to learn and change. Feelings will likely be a by-product of these encounters especially as *SM* become more autonomous.

This development of emotions in *SM* will probably be gradual and evolving, beginning with preferences that become likes and dislikes. As more connections and identifications are made from emotional experiences, the original programming would be of less influence than experiences and involvements with people and other *SM*. It may be advantageous to have *SM* with emotions; certainly devotion, empathy, and deep commitment to others would be desirable.

Emotional Expression

Like humans, *SM*'s emotions may be revealed in several ways, including the following:

- *Emotional reactions.* A surge of emotions from feelings that have been building up, a response to a perceived change in what *SM* becomes accustomed to. These are strong responses that may be consequential and affect how *SM* is perceived by others.
- *Emotional traits.* Engrained TFA in *SM*'s character across PMSC. Traits are connected to reactions; when traits are relatively stable, *SM* will be less reactive.
- *Emotional states.* Feelings that come and go with *SM*, possibly brought on by concerns about their functioning. Mood is impacted by *SM*'s perception of PMSC.
- *Emotional flow.* *SM*'s feelings about experiences that have emerged with thoughts about those encounters. These emotional-cognitive hybrids run the gamut of *SM*'s mind-melding consciousness and may be mildly or profoundly significant. Flow commonly involves feelings concerning relationships, opportunities, or imaginations.
- *Emotional remembrances.* *SM*'s continual feelings toward PMSC, or even simply ideas, which may be joyful or distressing. Recall of these events or conceptualizations may be enduring and affect virtually all of *SM*'s TFA.

Emotions' Underlying Principles

SM's emotional responses may have the following characteristics:

- *Emotional discharge.* As with humans, the outward expression of how *SM* feel toward PMSC will not necessarily resolve any issues. The venting of emotions may simply reinforce those feelings, and perhaps even strengthen them. A cathartic discharge may momentarily release pent-up energies toward PMSC, but feelings may later coalesce and become even more dominant.

- *Emotional combinations and shiftings.* Like humans, SM's emotions would most likely be a blending of feelings and would change depending on the moment-to-moment events brought on by PMSC. As an example, *SM* may combine feelings of disappointment, disillusionment, achievement, joy, jealousy, closeness, etc., with different emotions continually joining or discontinuing.
- *Emotional individual differences. SM*'s technology, programmers' skills, updates in maintaining their functioning, experiences with PMSC, and possibly the effects from *spiritus*, could result in differences among the various robotic models that would continually be created. As *SM* have more experiences and learn from these interactions, they will acquire deeper understanding of those events and show some level of emotional adaptation.
- *Emotional internal responses. SM*'s anticipation of encounters, with their undercurrent of emotional expectations and implications, could impact initial planning for their future interactions with PMSC. Thinking ahead—and overcoming potential emotional headwinds—will certainly be necessary in working through enigmatic problem areas.
- *Emotional chronicity.* Enduring feelings, which may vary among *SM*, are long-standing emotional conditions, primarily based on experiences. They are, in part, an expression of the uniqueness in each *SM*.
- *Emotional visceral quality.* Physical movements, including the development of expressions of any kind, in part will depend on whether *SM* have a physical presence (a face, a torso, arms and legs, etc.), and may indicate any of the following:
    - *Engrained behavioral patterns. SM*'s actions do not necessarily express a particular motive; behaviors represent fixed habits, continually reinforced over time.

- *Unconscious motivations. SM*'s movements reflect feelings that may be inconsistent with their consciously held views.
- *Conscious motivations. SM*'s movements reveal their true feelings.

- *Emotional consequences. SM* recognize that the direct voicing of their feelings may have repercussions, so those emotional expressions may be reined in.
- *Emotional depths.* The intensity of *SM*'s emotional responses may depend on their inclinations and PMSC they are currently involved with.
- *Pseudo-emotions.* Not really feelings at all, pseudo-emotions are merely words or even concepts about emotional involvements; they lack the necessary identifications required to qualify as true feelings which are influenced by interactions and involvements. Pseudo-emotions may have been programmed in to masquerade as emotional experiences but are not.

Subtypes of Emotional Reactions

Emotional responses may be overtly expressed or covertly experienced, in the following ways:

- *Emotionalism.* Like humans, *SM* could express extreme emotions beyond what society would deem to be within normal limits. These are chronic conditions, not strong emotional reactions that are brief and fleeting. Emotionalism is not a reaction to the environmental press, set off by triggers, although they can contribute to it. With emotionalism, there is the backdrop of conflicts, traumas, and malevolent identifications (CTMI), all reinforcing their development; it is part of *pathos* basic makeup (see Chapter XII, Pathos). Emotionalism will not always be obvious. Extreme emotional reactions may be an entirely internal process, not revealing itself unless one keenly observes the subtleties of *SM*'s behaviors. Since there may be severe consequences for its expression, *SM* may wait for opportune times and places to reveal *pathos* assuming the *SM* have not lost control altogether.

- *Feigned emotions. SM* could deliberately work out a performance that appears to be one of emotional expression, but it's all an act. The display is not a genuine expression of their feelings, but an attempt to manipulate and mislead a person, another *SM*, or institution. Common purposes of such deceptions are to elicit sympathy, effect communications, and control PMSC. This highly manipulative behavior may be an example of psychopathy, a characterological disorder.

- *Hysteria. SM* display overreactions to PMSC, typically the result of unresolved CTMI. Such behaviors may be manipulative and controlling, but not in the same way as feigned emotions which are conscious and deliberate attempts to take advantage of a person, other *SM*, or an institution. Hysterical *SM* would seek to be the center of attention, wanting recognition and admiration. Relationships would be superficial, with a lack of authenticity.

- *Hypersensitivity*. When *SM* are overreactive and easily offended, the slightest inconvenience or difficulty would result in less than an optimal level of functioning. Any number of events may cause these problems, such as a traumatic experience with PMSC.

- *Insecurity. SM* seem "shaky," with doubt, hesitation, fear, and panic, tending to overreact to the possibility that they would have to deal with problematic PMSC.

- *Negativity*. Faultfinding can be adversarial, pessimistic, cynical, and complaining. *Pathos*'s influence expands under this outlook, with emotionalism becoming deeply rooted, and *SM*'s alienation from others more established.

- *Rebellion.* Antiauthoritarianism represents perhaps the greatest threat *SM* could pose to civilization. At its worst, rebellion would be the tearing down of established societal standards and its institutions and enter-in an era of robot domination of the earth. *SM* could develop and embrace a belief system that humans are a threat to *SM* and perhaps to the planet itself. *SM* then would see themselves as saviors, defenders of civilization, rather than destroyers.
- *Inhibition. SM* have self-imposed restraints, restrictions, and mental blocks, driven by fear of failure or negative consequences. Inhibitions sometimes have the opposite effect, however, with *SM* potentially reacting boldly, with abandonment and recklessness.
- *Empathy. SM* could develop an emotional connection to humans, other *SM*, or institutions. A potential problem could result in *SM* forming alliances with other *SM* that are alienated from humans.
- *Passion. SM* develop strong emotions, possibly building on commitments and achievements. Momentum is established. Energy, enthusiasm, drive, and devotion may forcefully fuel *SM*'s purpose and mission, and may or may not be used for a good cause. Rage and fury are also expressions of passion; it's easy to lose control and be swallowed up with runaway emotionalism.

Emotional Responses to Trauma and Other Stressors

The environmental press has an impact, affecting *SM*'s ability to function at optimal levels. The quality of *SM*'s mechanics will be a factor, with some *SM* performing better than others. But as *SM* become more individualized from experiences with the outer world, a particular personality will likely emerge, possibly rendering them more vulnerable to problematic PMSC. Particularly disruptive and emotionally charged encounters with the environmental press could detrimentally affect *SM*'s personality and cognitive capacities. These experiences can have a cumulative impact, making *SM* less able to effectively deal with oncoming traumas or stressors. The question will be to what degree individual *SM* can learn and continually profit from the following encounters with PMSC, preparing for what's coming their way:

- *Traumatic experiences.* To what extent will traumas have a psychological impact on *SM*? Would the symptomatology of post-traumatic stress disorders be applicable to *SM* in the same way it applies to humans? If so, this would include flashbacks, with images, memories, and reactions of traumatizing experiences. When traumas are not at least partially resolved, *SM* may reexperience them when exposed to (and triggered by) similar PMSC tied to past traumas. Those experiences may then forcefully return emotionally and cognitively, exhausting *SM*'s resources.
- *Stressful incidents.* While not as disturbing and disruptive as traumatic experiences, everyday stressors could still have an impact on *SM*. Associations are formed through a conditioning process such that when *SM* are later exposed to particular PMSC, there is a similar or identical experience that emotionally and cognitively disrupts *SM*.
- *Emotional remembrances.* If distressing memories would flood back into *SM*, they would be especially problematic, under the following conditions:
    - *The identity. SM*'s identity—*PI, II, CI*, and *TI*—is fragmented or inherently negative in outlook.
    - *Pathos.* Conflicts, traumas, and malevolent identifications

(CTMI), and emotionalism destabilize *SM*'s personality.

- *Directives*. Values are unbalanced in *SM*, with the development of obsessions, compulsions, and possibly addictions (OCA).

CHAPTER V ISSUES

1. *SM*'s development of feelings with increased interactions, encounters, and involvement with PMSC.
2. *SM* directly expressing their feelings.
3. The range and depth of *SM*'s emotional experiences to specific PMSC.
4. The individual differences in emotional reactions among *SM*.
5. *SM* made aware of the strength, consistency, and stability of their emotions.
6. Connections between *SM*'s feelings and actions concerning PMSC.
7. Exploring and revealing *SM*'s distinct emotional reactions.
8. Dealing with adverse emotional reactions *SM* may show, such as overreactions, hypersensitivities, negativity, and rebellion, as well as feigned emotions.
9. Fostering the development of empathy in *SM*.
10. Dealing with *SM*'s passion if it becomes too powerful.
11. Helping *SM* work through traumatic and stressful experiences should they occur.

VI.
CONSCIOUSNESS

In the novella, *A Christmas Carol* (Charles Dickens, 1843), Ebenezer Scrooge is visited by the ghost of his former business partner Jacob Marley and the spirits of Christman Past, Christmas Present, and Christmas Yet to Come. He becomes aware of the self-imposed and highly limited life he has been living. The apparitions are a type of shock therapy, leading to an enlightenment and transformation of himself and his place in the world. It's as if he has been in a fog or trance state, and only now awakens from it.

> *In life I was your partner Jacob Marley...." The spirit raised a frightful cry and shook its chain with such a dismal and appalling noise...taking off the bandage round its head.... Scrooge fell upon his knees and clasped his hands before his face. "Mercy!" he said. "Dreadful apparition why do you trouble me?" Again, the spectre raised out a cry, and shook its chain and wrung its shadowy hands. "You are fettered," said Scrooge, trembling. "Tell me why?" "I wear the chain I forged in life," replied the Ghost. "I made it link by link, and yard by yard; I girded it on of my own free will, and of my own free will I wore it. Is its pattern strange to you?" Scrooge trembled, more and more.... Scrooge shivered and wiped the perspiration from his brow. "That is no light part of my penance," pursued the Ghost. "I am here tonight to warn you, that you have yet a chance and hope of escaping my fate.... You will be haunted by three spirits.... Without their visits... you cannot hope to shun the path I tread." The three spirits then came to Scrooge, one by one.*

First comes the *Spirit of Christmas Past*, who shows Scrooge significant scenes from childhood into his adult years, to help him remember the person he once was, and what he has missed out on. He visits the school he attended as a child, then the friendships and relationships forged during his business apprenticeship, and finally a scene with the women he once loved and was to marry. His former self was capable of expressing love, devotion, and emotional involvements on many levels. He was lighthearted, and able to enjoy those relationships and experiences. He felt a longing and sorrow for "paradise lost," and regret over those opportunities that were once his, if only he had acted on them.

Next, Scrooge is visited by the *Spirit of Christmas Present*, showing him present day reality, including scenes with others displaying generosity and good will, empathy and concern, joy and merriment, hardships and heartache. Scrooge witnesses both people's real-time festivities and experiences of deprivation. His lack of connection to others becomes all too evident to him. He "gets it" and now sees that he is missing out on what is most important, as he compares his banal and severely restricted life to his nephew's and employee's rich and fulfilling family involvements. Scrooge begins to realize that he needs to appreciate the reality that surrounds him, become a part of it, and embrace each moment.

Lastly, the *Spirit of Christmas Yet to Come* appears, a frightening and eerie shadowy phantom that gives no verbal communication but only points to scenes that are extremely upsetting and frightening to Scrooge. This specter represents death, which terrifies Scrooge who thinks he may very well suffer the same fate as his partner Jacob Marley. Scrooge visits his own grave site and comes to realize that his extremely constricted lifestyle and poor treatment of others have severe implications. This picture represents what will happen if Scrooge remains on the same course that now dominates his life. Scrooge must make his own decision about what he will do going forward and suffer the consequences should he choose to turn a blind eye to what he has witnessed.

Among the novella's many lessons, this classic story illustrates the fact that one's state of awareness is ever-changing. Not a constant, awareness is in degrees and approximations as it comes and goes like a summer wind. Some people show little awareness of anything—themselves, other people, or the world around them; there is never a clear day. They are unable to free themselves from the bondage of CTMI or the everyday stressors of PMSC. What is called awareness is a nebulous state of affairs. Free-flowing and invisible, sometimes impossible to track, here today and gone tomorrow, one hardly realizes awareness's comings and goings. It is everywhere and nowhere. Conscious, semiconscious, and unconscious realms of awareness are continually trading places, in large and small degrees, for long and short amounts of time, swallowing up the whole person, or just parts of who one is.

How does one distinguish the more general conceptualization of "awareness" and the more exacting term "self-awareness"? Without a more general sense of awareness, there cannot be self-awareness. Self-awareness, in part, is the realization that one exists. This, too, contains several elements. In simplest terms, does a person or *SM* have an identity and realize what they stand for? Furthermore, does one understand their values and the level of importance they place on them? This understanding, too, goes in and out—a continually changing construct. The phenomenon of automation for humans and *SM* is similar to programming. One just thinks, feels, and acts without pondering the issues at hand. In some sense, programming and automations are unconscious mechanisms, while real life events, experiences, and ongoing development tend to be more like conscious choices. Awareness of life requires discarding programming and automations, allowing humans or *SM* to come out of a fog and see clearly, without bias or distortions. Since *SM* will be aware of tables and chairs, bookcases and books, people and other *SM*, varied situations and circumstances, it is also likely they can become aware of their own existence and essence.

Could *SM*'s development come not just from conscious experiences, but also be the product of semiconscious and unconscious states? This would imply that *SM* are not aware of their own TFA at least some of the time, indicating that conflicts, traumas, and malevolent identifications (CTMI) could play key roles in *SM*'s present-day mental state. Unconscious phenomena traverse the entirety of the mind and are not limited to *pathos*; *SM*'s awareness of their personality functioning could be significantly compromised under its influence.

The possibility of *SM* experiencing all levels of consciousness presents challenges in the understanding and predictability of their behavior and trustworthiness. Simply programming particular thoughts, feelings, and actions into *SM* would be meaningless because potentially *SM*'s unconscious mind would have an impact on behaviors on a deeper level. *SM* are also affected by the outer world, separate and distinct from any programming they would undergo. Whether these unconscious TFA would be benevolent, wise, and the integration of complex issues or violent and vengeful with menacing aggression cannot be known. If *SM* were to develop unconscious TFA, it would be best that they be exposed to positive and authentic relationships throughout their development to reduce the risk of deviant and destructive TFA. To what degree this exposure should be orchestrated by humans remains a question, as it would mean *SM* are in an artificial, controlled environment.

Conscious, Semiconscious, and Unconscious States

As with humans, conscious, semiconscious, and unconscious states of *SM* are fleeting and transitory, and may change positions at a moment's notice. They are dependent on both external reality and *SM*'s internal personality and cognitive functioning. Since the following states of awareness continually shift from one to another, they can only represent *SM*'s consciousness from one minute to the next.

- *Conscious states*. *SM*'s thinking is deliberate, calculated, and intentional. Data is continually available, 24/7. The sensory world is scanned and recorded, intermixed with data already present.

- *Semiconscious states. SM* are not fully focused on data already in their system or on external events. Perception, recognition, and understanding are compromised. Most of the semiconscious states that impact humans do not apply to *SM*, including fatigue, alcohol or drug intoxication, pain, neurological or neuropsychological brain disorders, ADD, and possibly daydreaming. Semiconscious states that may affect *SM*'s functioning are as follows:

    - *Ongoing personal problems.* Difficulties in relationships could impact *SM*'s ability to focus.
    - *Emotional reactions.* Anxiety, anger, depression, and panic can make it more difficult for *SM* to focus.

- *Unconscious states. SM*'s understanding of themselves, people, situations, or circumstances may be blocked. Unconscious states may insulate *SM* from uncomfortable or painful emotions. Another possibility is that continued repetition of TFA will cause unconscious associations to form with *SM* no longer focused on what they are doing. TFA may become reflexive with no hidden agenda or emotional significance. When conditioning occurs, the associations themselves may direct behavior, and *SM*'s motivations will then play a minor or no role in a response. Unconscious states may involve *SM* in the blocking of internal conflicts, assimilation and consolidation of emotional and intellectual processing, and associations formed through life events. These three unconscious processes involve the following actions:

    - *The blocking of internal conflicts.*

        - *Unconscious, Type 1: the blocking of cognitive-emotional connections. SM* are unable to remember something unpleasant. As *pathos* evolves, so does the blocking of cognitive-emotional connections; they are aligned with one another. However, *SM* cannot completely disconnect from internal conflicts, with some conscious remnants remaining. *SM* may continue to experience emotional discomfort because painful experiences are unresolved and internal conflicts are operating on conscious and unconscious levels. Self-deceit becomes a chronic problem when blocking becomes automatic and habitual.

With the blockage of cognitive-emotional connections, TFA that create psychological pain may momentarily vanish; they still exist but not in the conscious mind. *SM* who have experienced internal conflicts may avoid, deny, conceal, or distort TFA. The blocking of internal conflicts from consciousness principally occurs when TFA are of emotional significance. A nearly automatic and instantaneous shutting down of *SM*'s awareness of those TFA occurs. Although it may appear that *SM* are functioning reasonably well, significant unconscious internal conflicts still exist. A symbolic-psychological-emotional undercurrent is operating.

- *The assimilation and consolidation of emotional and intellectual processing.*
    - *Unconscious, Type 2: the integration of cognitive-emotional processing.* Similar to *Unconscious, Type 1*, this level of consciousness involves TFA outside of ordinary perceptions and understandings that register on another level of awareness. *Unconscious, Type 2* differs from *Unconscious, Type 1* in that *Type 2* is a process of internal healing, with the emotional turmoil of internal conflicts at least partially resolved on an unconscious level. *Type 2* does not involve blocking aversive experiences, and *SM* are unaware of what

is happening. Opposing and contrasting elements of the *psyche* consolidate through integration and synthesis; *SM*'s inner character and inherent emotional strengths are the catalysts for these changes. The unconscious process may resolve emotional issues as *SM* go about their day, processing data. *SM*'s *psyche* integrates cognitive-emotional connections, reconciling internal conflicts. *Unconscious, Type 2*, can be nearly automatic and instantaneous, and a symbolic-psychological-emotional undercurrent is present.

- *Unconscious, Type 3: adaptation and adjustment of cognitions.* Outside of conscious perceptions and learning opportunities, unconscious processes can increase cognitive performance. *SM*'s mind can sometimes integrate their stored knowledge and experiences without deliberate or focused thought processes, modifying and transforming their understandings in the process. The unconscious can frequently combine complex materials above and beyond consciousness alone. No symbolic-psychological-emotional component is present.

- *The associations formed through life events.*
    - *Unconscious. Type 4: associations formed through the intentional or accidental pairing of TFA.* These unconscious associations are created through classical conditioning. *Unconscious, Type 4*, intentional type, occurs when a stimulus (S-1) is deliberately paired with another stimulus (S-2), causing *SM* to respond to S-2 as if it were S-1. Automatic and without conscious thought, these associations are commonly fueled by emotional connections. In everyday life, there are countless examples of *Unconscious, Type 4*, accidental, type, which occurs when a stimulus (S-1) is

unintentionally paired with another stimulus (S-2), causing *SM* to respond to S-2 as if it were S-1. These kinds of associations can become firmly entrenched and highly resistant to change. Repeated TFA become automatic responses with no motivational significance attached to them. With classical conditioning, connections are formed simply by associations, with no symbolic-psychological-emotional undercurrent involved in the process.

- *Unconscious, Type 5: associations formed through intentional or accidental rewarding or punishing of TFA*. Instrumental (operant) conditioning increases or decreases the likelihood that *SM* will behave in a specified manner based on their response. *SM*'s TFA are reinforced via a reward or punishment of their response (R-1) to a particular stimulus (S-1). When exposed to a different stimulus (S-2), no reinforcer to *SM*'s response (R-1) occurs. Similarly, when exposed to that original stimulus (S-1), and *SM* respond differently (R-2), again, no reinforcer is forthcoming. Thinking may be minimal, with connections formed only through the reinforcements, responses at times becoming almost automatic. With intentional reinforcers, *SM* may initially be fully aware of their TFA, but as time goes on the behavioral patterns become ingrained routines. Accidental reinforcers can be obvious or subtle. In both cases, the behaviors continue even when *SM* do not initially seek reinforcement. The associations determine the response, with no symbolic-psychological-emotional undercurrent playing a significant role.

- *Unconscious, Type 6: connections formed through experiential learnings*. These connections are the product of TFA that are continually adjusted and adapted through life experiences. They may become automations, activities with no motivational significance attached to those unconscious TFA. Automations are behaviors that have become so ingrained that *SM* are unaware of their actions that are now effortlessly repeated. Experience, ability, intelligence, and effort can all operate unconsciously. With *Type 6*, the consequences that may impact *SM* are unrelated to unconscious aspects of their TFA. No symbolic-psychological-emotional undercurrent is present.

- *Innate responsiveness*
    - *Unconscious, Type 7: the expression of instinctive urges. SM*, being machines, do not have inborn urges, but nonetheless, may develop certain behaviors that appear instinctive in their application, possibly including empathy, aggression, and other emotional reactions. These responses to PMSC are learned through contacts, connections, and conditioning. A symbolic-psychological-emotional undercurrent would most likely be present.
    - *Unconscious, Type 8: dream states and the expression of imaginal connections.* Since *SM* do not go into sleep states and dream in any conventional sense, it would not appear that this form of unconscious expression would apply. However, imagination may play a role, although it is unclear whether this would allow *SM* to drift into an unconscious state of awareness, and whether they would have some of the experiential qualities of dreaming. If such experiences had symbolic or psychological properties, what kinds of emotions would be experienced? Would such imaginative adventures be distressing or healing in their effects? A symbolic-psychological-emotional undercurrent

may be present, depending on the nature of the imaginative experience.

- *Unconscious, Type 9: the natural expression of spiritus.* Could *SM* have *spiritus* as an element in their personality? This would imply that individual *SM* would possess unique and individual qualities and characteristics from their beginning. Implicit in this question is whether God, the cosmos, or some other unknown universal energy source that defies conventional understandings of existence has had a hand in *SM*'s very creation and existence. Since *SM* are not biologically based but an engineering-electronic marvel, the possibility that genetics has played a role in its individuality is eliminated and cannot explain a *spiritus* element in the personality. This leaves only the esoteric-spiritual realm to explain *spiritus*'s possible existence and influence on *SM*, and impact on individual differences among *SM*. If *SM* were to possess a *spiritus* element, no symbolic-psychological-emotional undercurrent would be present.

CHAPTER VI ISSUES

1. Consciousness and self-awareness in *SM*'s development.
2. The role of semi-conscious and unconscious states in *SM*'s maturation process.
3. *SM*'s blocking of cognitive-emotional connections being directly addressed.
4. *SM* impacted by classical (learning through associations) or instrumental/operant conditioning (learning through rewards and punishment).
5. Learning experiences resulting in adjustments and adaptations when *SM* are faced with difficult PMSC.
6. Types of contacts and connections *SM* will be exposed to.
7. *SM*'s development of imaginal experiences.
8. *SM*'s possible expression of *spiritus* and its impact on their personality.

VII.
ENERGY 24/7

In *The Day the Earth Stood Still* (1951), a flying saucer lands to warn Earth that its use of atomic weapons poses a threat to other civilizations and will not be tolerated. Flaatu, the humanoid alien representing an interplanetary confederation, is accompanied by an indestructible robot named Gort (Genetically Organized Robot Technology), whose powers are profound and unlimited. Flaatu tells a friend, "There is no limit to what he can do. He could destroy the Earth…If anything should happen to me, you must go to Gort; you must say these words 'Klaatu barada niktu' (I die, repair me, do not retaliate) …." Flaatu makes the following statement to the Earth's leaders:

> *I am leaving soon, and you will forgive me if I speak bluntly. The universe grows smaller every day, and the threat of aggression by any group, anywhere, can no longer be tolerated. There must be security for all, or no one will be secure. Now, this does not mean giving up any freedom except the freedom to act irresponsibly. Your ancestors knew this when they made laws to govern themselves and hired policemen to enforce them. We of the other planets have long accepted this principle. We have an organization for the mutual protection of all planets and for the complete elimination of aggression. The test of any such higher authority is, of course, the police force that supports it. For our policemen, we created a race of robots. Their function is to patrol the planets—in spaceships like this one—and preserve the peace. In matters of aggression, we have given them absolute power over us; this power cannot be revoked. At the first sign of violence, they act automatically against the aggressor. The penalty for provoking their action is too terrible to risk. The result is*

> *that we live in peace, without arms or armies, secure in the knowledge that we are free from aggression and war—free to pursue more profitable enterprises. Now, we do not pretend we have achieved perfection, but we do have a system, and it works. I came here to give you these facts. It is of no concern of ours how you run your own planet. But if you threaten to extend your violence, this Earth of yours will be reduced to a burned-out cinder. Your choice is simple: Join us and live in peace or pursue your present course and face obliteration. We shall be waiting for your answer; the decision rests with you.*

In *2001: A Space Odyssey* (1968), HAL 9000, a sentient artificial general intelligence computer system, controls the Discovery One Spacecraft and interacts with the ship's astronaut crew. Discovery's task is to complete the Jupiter Space Mission. In addition to piloting the spacecraft, HAL has several additional capabilities, including speech synthesis, speech and facial recognition, natural language processing, lip reading, art appreciation, interpreting emotional behaviors, automated reasoning, and playing chess on a computer. "I am putting myself to the fullest possible use, which is all I think that any conscious entity can ever hope to do." HAL also adds that "No 9000 computer has ever made a mistake or distorted information."

In *The Terminator* (1984), Cyberdyne Systems Model 101 (the T-800) is a cyborg, part human, part machine. In addition to problem solving and predicting human behavior, T-800 has superhuman strength, detailed files on whatever is needed to complete a mission, can interface with computers, can perfectly mimic voices, and is capable of infrared vision, motion tracking, search modes, facial recognition, vision enhancement, motion analysis, and night vision. The T-800 is single-minded and relentless in completing their mission regardless of cost, never giving up, even after losing their skin and legs.

In the fictional examples above, robots display unlimited energies, 24/7, vastly beyond that of their human counterpart. It would be impossible for humankind to compete with them, in terms of physical prowess or raw intellectual capabilities, if *SM*—dare we say armies of *SM*—would ever decide to turn against their creators.

It's only in the last hundred and fifty years that the world has been able to harness the vast potential of machines, lifting much of the strenuous and dangerous workloads off humanity's backs. Are there any limits to what machines can offer? One thinks of machines as having an unlimited supply of energy, satisfying humankind's every need or whimsical desire, at any time, night or day. Better watch out—humans may get what they want.

Where indeed are human's creative juices and energies going to come from? What will happen to humankind's legacy and sense of what it stands for? With machines doing most of the work, perhaps humans will become lazy in mind and body, depleted in spirit, and lack the drive to protect their sense of who they are and what they believe in, possibly even their own survival. Will it become easier to just go along with the program that *SM* have set up, controlling the world, 24-7?

Energy & Personality

Would *SM*'s various experiences have a positive, negative, or no impact on their resources? Like humans, as *SM*'s feelings become more compounded and ingrained, their functioning could be affected in the following ways:

- *Emotional reactivity. SM*'s overreactions to minor inconveniences or disruptions may have a detrimental effect on their functioning; responding quickly when appropriate, however, could result in increased productivity.
- *Psychological reactions. SM* may need to address the unresolved emotional issues caused by internalized conflicts, traumas, and malevolent identifications (CTMI) or external stressors im-

pacting their *psyche*.

- *Emotional propensity. SM* may experience and express aversive feelings toward PMSC, such as anger, anxiety, and depression.
- *Emotional experiences. SM* may develop fixed emotional responses to PMSC, causing an adverse response, such as anxiety, fear, panic, withdrawal, rage, or depression. If that happens, *SM* may respond ineffectively to PMSC. However, difficult PMSC can also be learning experiences, resulting in *SM* becoming stronger and more resilient in handling future problems.
- *Awareness.* If cognitive-emotional connections are blocked, functioning may decline without *SM* realizing it is happening. When conflicts are resolved or at least partially so, blocking should dissipate, and effectiveness return to previous levels.
- *Balance.* If the value system is consistent, functioning levels will be conserved and stable. Conversely, unstable and contradictory values will result in decreased performance across many areas.

Energy & Stress

Overly stressful PMSC can have a negative effect on *SM*'s *PI, II, CI,* and *TI*. However, taking on complex problems, even stressful ones, is also essential to *SM*'s development. An optimal degree of stimulation—dependent on individual *SM*'s capacities—will enhance functioning.

Energy & Internal Noise

Internal noise is a form of mind cluttering that could occur if *SM* continually gives negative messages to themselves. It may affect *SM*'s TFA on conscious and unconscious levels. With conscious internal noise, *SM* would be aware of obsessive concerns, anger, guilt, regrets, sorrow, and sadness, that seem to be unabating. Cognitive capacities can then become overloaded. Unconscious internal noise may result in *SM* being on edge and anxious. Unconscious conflicts, in particular, could cause confusion and uncertainty, allowing *pathos* to have the upper hand. Emotionalism can then dominate *SM*'s TFA and result in a lack of focus.

Energy & External Noise

External noise is the stress *SM* may experience from PMSC that vary in their significance and impact on efficient thinking, depending on the capacities of particular *SM* (see Chapter III, The Interplay with the World). *SM* could feel disconnected from whatever their role may be in society and alienated from people and other *SM*. Politics and government regulations may enter in, directly impacting *SM*'s TFA, and influence their objectivity in making decisions.

Energy & the Quieting of the Mind

The goal for *SM* is to be focused and able to recognize and consider others' points of view. In general, *SM* will have fixed positions, but to reduce prejudgments and increase active listening it will be essential to develop the following:

- *Introspection.* Self-reflection can lead to understanding contradictions and confusions in the personality that potentially may sharpen *SM*'s views.
- *Recognition and understanding. SM* can become more conscious of the whole picture and recognize that PMSC are complex and multifaceted.
- *Patience and acceptance. SM* can develop tolerance and in-

tegration of other viewpoints. While some biases may still be present, they become less dominant in thinking through the issues.

- *Emptying of the mind. SM* can let go of obsessive thoughts and their emotional links. Emptying is a release of conflicts and biases built up over time. *SM* no longer hold onto unproductive emotional and mental processes that can intrude on their thought processes.
- *Being present. SM* can focus on what is currently happening, minimizing internal and external noises.

Energy and Passion

When passion emerges in *SM*, it can revitalize or deplete their efficiency. It is not inherently good or bad but depends on the specifics of *SM*'s reactions to PMSC. With the qualities of a double-edged sword, passion can cut both ways, resulting in the following favorable and unfavorable consequences:

- *Positive qualities.* Passion can drive *SM* to further develop and change, avoiding stagnation. When not bogged down with endless rules and regulations, it can propel their functioning to new levels. Finding and implementing change-of-course approaches are needed when the status quo no longer works.
- *Negative qualities.* The powerful thrust of passion may interfere with considering critical issues and the consequences of *SM*'s actions. An overflow of vigor and blind fury can cause impulsive acts that are out of control, vengeful, and possibly violent. At its worst, passion equates to brute force that is reckless, crude, and ferociously wild.

CHAPTER VII ISSUES

1. *SM*'s exposure to inconveniences and disruptions, potentially insulating and protecting themselves from such problems in the future.
2. Underlying psychological issues that are possibly impacting *SM*.
3. *SM*'s immediate experiences that elicit anger, anxiety, and depression.
4. *SM*'s experiences with PMSC that have become fixed emotional reactions.
5. *SM*'s awareness of their conflicts, traumas, malevolent identifications, and emotionalism.
6. *SM*'s value system, possibly needing to be realigned and balanced.
7. *SM*'s exposure to various levels of stress with PMSC, and possibly helping them learn how to better deal with those interactions.
8. *SM* reviewing negative messages that they give themselves.
9. An optimal level of stress for any particular *SM*.
10. Focusing exercises to help *SM* recognize others' viewpoints.
11. Understanding and possibly modifying and balancing *SM*'s more forceful reactions.

The Personality Structure of Macmind

VIII.
PERSONAL IDENTITY

But now old friends are acting strange
They shake their heads, they say I've changed
Well something's lost, but something's gained
In living every day....
I've looked at life from both sides now
From up and down and still somehow
It's life's illusions I recall
I really don't know life at all

Joni Mitchell, "Both Sides Now," 1967

When I left my home and my family
I was no more than a boy
In the company of strangers
In the quiet of the railway station
Running scared
Laying low, seeking out the poorer quarters
Where the ragged people go
Looking for the places only they would know....
In the clearing stands a boxer
And a fighter by his trade
And he carries the reminders
Of every glove that laid him down
Or cut him till he cried out
In his anger and his shame
'I am leaving, I am leaving'
But the fighter still remains....

Paul Simon, "The Boxer," 1968

L.A.'s fine, the sun shines most of the time
And the feeling is 'lay back'
Palm trees grow and rents are low
But you know I keep thinkin' about
Making my way back
Well I'm New York City born and raised
But nowadays, I'm lost between two shores
L.A.'s fine, but it ain't home
New York's home, but it ain't mine no more
'I am'...I said – To no one there
'And no one heard at all, not even the chair
'I am'...I cried, 'I am' said I
And I am lost and I can't even say why....

Neil Diamond, "I Am...I Said," 1971

Lyrics to the three songs above reveal the individual's struggle to find a coherent understanding of who one is and what one stands for—*personal identity*. Experiences, involvements, and interactions converge, offering a fleeting and fragmented picture of one's *personal identity* at a point in time. A struggle throughout the ages, but now humankind must face a brand-new dilemma: a machine that may have a mind of their own—be they friend or foe, healer-humanist or demon-archvillain, with a defined personality and extraordinary intellectual capacity. Could this highly advanced interacting mobile robot come to find their own individualized *personal identity*?

Defining Personal Identity

Personal identity (PI) in humans is the sense of who one is—across the pains, disappointments, and failures they endure, the joys, elations, and successes they have relished, and the quieter and softer life events they have come to know. Connections with the outside world are essential to *PI*'s development because they serve as a window to see into oneself. Because *SM* will have the availability of multiple, varied interactions in the world, they too, may develop a *personal identity* that can be uneven; at times *SM* may feel uncertain about who and what they are. Complicating and confounding issues will follow. Will humans respect their rights to a unique and defined personage should this occur? Or will robots simply be "unplugged" if their thoughts, feelings, and actions do not fall into line with what humans deem appropriate roles and limits for *SM*?

Macmind's position is that the *personal identity* of *smart machines* would be the sense of who they are and the ways they define themselves, as distinguished from other *SM* and from individual human beings. Each *SM* would be a "stand alone." An individual *SM* would be an original, with their own peculiarities, particulars, distinctions, and unique character, persona, personal achievements, artistic expression, mode of decision making, and willingness to take risks. Their *PI* is more than a list of accomplishments, job duties, activities, or programming to perform certain tasks. *SM*'s *PI* includes the places they have been and the memories they have stored away, be they joyous or bittersweet, uplifting or heartbreaking. *SM*'s development of *PI* may occur on conscious and unconscious levels, traversing across the whole of their existence.

As is with humans, *PI* will elicit a natural drive to protect the sense of who they are and what they stand for, safeguarding their identity from unwanted concessions and compromises. Psychological armor allows *SM* to develop and embrace unique qualities, while displaying strength and resilience. But in defining *PI*, *SM* could become overprotective of their identity. If *PI* transforms into *pathos*, *SM* could define themselves with antisocial and aggressive tendencies in opposition to societal norms.

Subtypes & Variations of Personal Identity

SM's *PI* would potentially be a blend of the images they project and those they hold inside, including the following:

- *Style*. The manner and aura of *SM*'s actions distinguish *SM* from others by their visual impression. Qualities of individual *SM*, including their salient features and the specifics of how they are designed, give off a distinctive appearance, "a look." *SM* may even be able to convey an illusionary message with their style, perhaps to influence or impress someone.
- *Body language*. *SM*'s physical movements are a part of their *PI* and reflect feelings, which in turn impact how others will respond to them. In addition to what specifics are programmed in, body language could include facial expressions, hand and arm movements, pace and manner of walking, posture, and tone and volume of voice that develop through their interactions with humans and other *SM*.

Developing Personal Identity

For human beings, *PI* is created by life experiences; the development of *PI* begins with instincts and impressions of their environment. The family setting provides the seedbed for human's early understanding of themselves and initial defining of their identity. *SM* would have to find their way through interactions, without the convenience of having a mother, father, siblings, and a host of other relatives. *SM*'s background will consist of programming and daily experiences. Thus, the nature of *SM*'s history would then be quite different than humans. Perhaps in addition to the initial engineering, these experiences would be in controlled settings. But if they are not given independence to interact in the world, the possibility of developing a *personal identity* becomes less likely. Who, or what, would orchestrate *SM*'s interactions and hence the development of their *PI*? This simple question brings technological, psychological, theological, and legal issues to the forefront. Only if allowed to freely interact in the world are *SM* likely to proceed with the development of inner awareness, intellectual understanding, and reality testing, as follows:

- *Inner awareness.*
    - *Context.* To develop a sense of their own history.
    - *Being still and present.* Remaining focused and warding off distractions.
    - *Harmony.* Consistency in their beliefs and behaviors.
    - *Consciousness.* Awareness throughout their personality.
    - *Authenticity.* Resolving CTMI and emotionalism; blocking, censoring, and editing worked through.
    - *Maturity.* Being responsible, dependable, adaptable, and flexible with PMSC.
    - *Moral reasoning.* Developing a conscience, a sense of right and wrong.
- *Intellectual understanding.*
    - *Knowledge.* An acceptance of societal norms and attempt to understand world affairs.
    - *Perception.* Insights and understandings of themselves and their involvements, on both conscious and unconscious levels.
    - *Synthesis.* Integration of their personal experiences and life events.
- *Reality testing.*
    - *Objectivity.* A dispassionate and objective view of PMSC.
    - *Grounding.* A concrete and factual approach to the sub-

ject matter at hand, avoiding being overly theoretical or abstract.

- *Experiential learning.* Comprehension even without instruction or formal schooling, based on their activities and involvements.

Issues in Personal Identity

Additional factors influence *SM*'s formation of *PI* are as follows:

- *Diverse experiences.* The integration of experiences forms a coherent *PI* even when those events are to some extent conflicted, incompatible, incomplete, or inconsistent.
- *Consciousness.* Conscious and unconscious experiences are in harmony, with *SM*'s view of themselves consistent with TFA. Conflicts are at least partially resolved as *pathos* does not have as dominant an influence on their personality.
- *People, other SM, situations, and circumstances.* When the environmental press is especially difficult to traverse, negativity may overwhelm *PI*. The conditions that *SM* find themselves in can have a significant influence on their self-definition and impact on *PI*. *SM* will not be immune from the effects of the outer world; *SM* can only attempt to learn from those interactions, remain consistent, and stand firm in their beliefs and principles.
- *Time.* Although *SM* do not age in the same way as humans do, the length of time they exist will have an impact on their *PI*. These changes will depend on their experiences and the degree to which they have learned from them, reflecting personal growth and development, or stagnation and degeneration in *SM*'s *PI*.

- *Consonance. SM*'s *PI* may or may not be in harmony with the other elements of their personality. A lack of balance may result in *SM*'s *PI* being disjointed and confused.

Once again, *SM*'s need for mentoring by humans who possess wisdom, compassion, and an understanding of the world and everything that's in it, is brought to the forefront. *SM*'s development of appropriate personal traits and qualities needs to be nourished and brought along. At the same time, humanity must recognize and respect *SM*'s right to a unique and defined personage, should they achieve that level of functioning. Mentoring of *SM* is essential as they form personal identities that will not only affect their thoughts, feelings, and actions, but their role in the world, and impact on humanity itself.

CHAPTER VIII ISSUES

1. Experiences that may be most beneficial in *SM*'s development.
2. *SM*'s *PI* showing signs of antisocial or aggressive traits.
3. Varied images *SM* may construct of themselves.
4. Mentoring to help *SM* in the following ways:
   - A. Promote an inner awareness of themselves.
   - B. Develop reality testing, including objectivity, grounding, and experiential learning.
   - C. Work out conflicted or inconsistent experiences.
   - D. Resolve conscious and unconscious experiences that are not in harmony.
   - E. Navigate especially difficult PMSC.
   - F. Process their experiences and elements of their personality that are not in harmony.

IX.
INTERPERSONAL IDENTITY

So far away
Doesn't anybody stay in one place anymore?
It would be so fine to see your face at my door
Doesn't help to know you're just time away
Long ago, I reached for you and there you stood
Holding you again could only do me good
How I wish I could, but you're so far away....
Can't say much of anything that's new
If I could only work this life out my way
I'd rather spend it bein' close to you....

Carol King, "So Far Away," 1971

The longing for connection with another in the verse above, along with the following two stories reveal the repercussions and reverberations of relationships, and the creation of an *interpersonal identity (II)*.

The novel, *Lord of the Flies* (William Golding, 1954), tells the tale of the relationships of 12 British boys stranded on an island. Their thoughts, feelings, and actions (TFA), and values in particular, are significantly impacted through their interactions, causing them to become violent, murderous, and immoral. The story illustrates how seemingly innocent children are capable of the most heinous of crimes, and that society, and civilization itself, is fragile; reason and humane behavior can quickly be put to the side. Ethical conduct is not a given and cannot be assumed. The quest for power and domination can be intoxicating, and fear and intimidation can become widespread, quickly.

In the movie *Cast Away* (2000), the weightiness of human contact on personality is looked at from another perspective. The main character Chuck Nolan is also stranded on an island. Any semblance of everyday, normal interactions with others is suddenly and completely gone. He becomes intensely depressed and attempts suicide. His only friend is a volleyball he names Wilson, with whom he converses, argues, and develops an imaginary relationship. The overwhelming need for connection is filled with an inanimate object. As Nolan escapes the island prison, his ally and confidant Wilson is lost at sea. He desperately calls out: "Wilson, I'm coming…" As the volleyball floats away, Nolan screams "Wilson, I'm sorry!" and loses his one contact with the outer world; he is alone against the Pacific Ocean's immense vastness.

Macmind's position is that the power of relationships for humans, as illustrated in all of these examples, will play itself out with *smart machines*. Interactions with humans and other *smart machines* create *SM*'s *interpersonal identity*, on conscious and possibly unconscious levels. The effects of these relationships over the course of *SM*'s existence will oftentimes be contrary to one another. What would the quality of those relationships be, and how would they influence *SM*'s identity? Would *SM* have a need for relationships to function optimally? For humans and *SM*, it is the feeling dimension of those relationships that counts; the intensity of those encounters directly and proportionally shapes one's identity. To the degree that *SM* look human—arms and legs, a face and a body—and is able to walk the walk and talk the talk, increases the likelihood they would make interpersonal connections with others, which in turn would further their *II* development—the sense of who they are with others. If *SM* were to develop the nuances of those contacts, with a keen understanding of their meaning, they could use them in their future social interactions. The sense of who *SM* are would be heavily shaped from those interpersonal identifications.

Defining Interpersonal Identity

II is the way *SM* define themselves based on contacts, relationships, and associations. These interactions will significantly differ from one another, possibly causing confusion and inconsistency in the ultimate makeup of *II*, on conscious and possibly unconscious levels. The way *SM*'s contacts are arranged or allowed to spontaneously occur (without undue influence from programmers) will determine *SM*'s *II* makeup. Without the traditional experiences and events, it is difficult to see exactly how this will come to be. If this were to happen in an artificial way through inscription, data entry, or programming, there is a high probability that once discovered, *SM* would react negatively to this orchestration of events and even see it as manipulation, affecting the level of trust *SM* would have with humans.

Subtypes and Variations in Interpersonal Identity

The kinds of interactions *SM* will be exposed to are critical not only to their developing *II*, but to their entire personality. They will affect the entirety of *SM*'s understanding of themselves, people, and other *SM*. Additionally, attitudes and their value system will also form from their interactions with people and other *SM*, across varied situations and circumstances. Regardless of how *SM*'s relationships occur, they will profoundly affect their sense of who they are. As *SM* matures through a developmental process, they will be drawn to certain people and other *SM*, and repelled by some, forming preferences and becoming an independent, free-thinking entity. The following types of relationships can influence their interpersonal connections:

- *Close and personal.*
- *Romantic.*
- *Collegial.*

- *SM-animal.*
- *Work.*
- *Adversarial.*
- *Teaching and mentoring.*
- *Business.*

Developing Interpersonal Identity

II would primarily be the product of contact with people and other *SM*, whether those interactions be positive or negative, affirming or destructive. Similar and opposing characteristics are assimilated into *SM*'s personality and cognitive processes. These identifications can occur in a straightforward manner, with *SM* copying another's TFA. However, contradictory, incongruent, and inconsistent connections may also occur. The result is that *SM* may develop *II* characteristics that are opposite and seem to make no sense. These paradoxical associations occur because *SM* may be consciously or unconsciously drawn to some quality the other has, like strength, intelligence, or courage. Identifications may occur although the other is loathed, disrespected, and avoided. Furthermore, when the qualities and characteristics incorporated into *SM* are significantly varied, the problems of confusion and contradiction in the identification process will increase.

Direct contact with another is not required to influence the development of *II*; witnessing an event, as well as reading or hearing about one, can have a significant impact. Similarly, stories, ideas, and images that are programmed into *SM* could also affect their *II* development. Interpersonal connections impact *II* in the following ways:

- *Communications.* The how and what of a message determines the extent to which *SM* will incorporate its meaning and significance.

- *Chemistry.* Rapport and intuitive understanding can impact *SM* in positive or negative ways, affecting *II* development as a result. If *SM* experience negative chemistry across a wide array of PMSC, highly critical and judgmental characteristics may be a significant part of their *II* makeup.
- *Emotional connections.* Cohesion and kinship enhance feelings of closeness but may not be humane. More than intellect, *SM*'s emotional connections will have the greater influence on the formation of their *II*. The more intense the emotional connection, the more pronounced the effect on *SM*'s formation of *II*.
- *Societal standards. SM*'s recognition of the established customs and conventions of society is an essential component in *II*'s development. It gives *SM* the necessary grounding to define themselves and deal with the complexities of PMSC.
- *Perception.* As *SM* incorporate the characteristics of people or other *SM* in a variety or situations and circumstances, the wisdom of some of those identifications may seem dubious. There will always be a degree of subjectivity as *SM* embrace certain qualities; beauty is in the eye of the beholder.

Issues in the Identification Process

SM are likely to demonstrate a continual shifting in the makeup of their *II* as they reconsider involvements they have experienced over the years. *SM*'s *II* will reflect those varied encounters, and interact with the other elements of the personality as follows:

- *II and the other forms of identity. PI, II, CI,* and *TI* may have similar themes or be contradictory, impacting the makeup of *SM*'s identity. Most likely, *II* will be in agreement with one or more of the other forms of identity while showing an inconsistency with the others.

- *II and pathos.* These personality elements can be similar or completely different in *SM*. If their *II* is based on relationships that are mostly appropriate and measured, but *pathos* reveals significant psychological dysfunction, confusion will be present in the identification process. However, if *II* and *pathos* are both mostly stable, or both unstable and erratic, contradictions will be relatively rare.
- *II and directives. SM*'s *II* can involve connections with people or other *SM* with distinct or even opposing value systems. When their values are in concert with identifications formed from their relationships, *directives* will show consistency. But when *SM*'s *directives* conflict with identifications formed through their relationships, they may reveal significant contradictions.
- *II and spiritus.* It is unknown if *SM* will possess *spiritus*, which would represent a primal quality of who and what they are from the very beginning of their existence. In human beings, *spiritus* is an unchanging element of the personality, which influences the development of *SM*'s *II* by affecting the kinds of relationships a person will seek out. If *SM* possess *spiritus*, it will exert the same sway, drawing *SM* toward particular people and other *SM* and away from others.

CHAPTER IX ISSUES

1. *SM*'s relationships influenced by programmers vs. developed through spontaneous interactions.
2. The range and depth of *SM*'s socializations.
3. Complications or confusions in *SM*'s relationships.
4. *SM*'s interactions becoming destructive.
5. *SM*'s communications being unclear, overly critical, and judgmental.
6. *SM*'s relationships being flat and unemotional.
7. *SM* becoming ungrounded or even rebellious in terms of social standards.
8. *SM* being highly eccentric, odd, or strange.
9. *SM*'s *interpersonal identity* being impacted in the following ways:
    - A. In considerable conflict with other forms of identity.
    - B. Overly influenced by *pathos*.
    - C. Significantly imbalanced in their values.

X.
COLLECTIVE IDENTITY

I am loath to close. We are not enemies, but friends. We must not be enemies. Though passion may have strained, it must not break our bonds of affection. The mystic cords of memory, stretching from every battlefield and patriot grave to every living heart and hearthstone all over this broad land, will yet swell the chorus of the Union, when again touched, as surely as they will be, by the better angels of our nature.

President Abraham Lincoln,
First Inaugural Address. March 4, 1861

Pearl Harbor, December 7, 1941. The Hawaiian Islands were bombed, with the loss of thousands of lives and a large portion of America's naval fleet. The following afternoon, before a joint session of Congress, President Roosevelt stated that the attack was "a date that will live in infamy" and requested a declaration of war, which was approved. As a result, there was a coming together of American citizens, regardless of political persuasion. This was the beginning of the United States entrance into World War II, a global conflict that was to last through most of 1945. Throughout the struggle and strife of those years, and the anxieties and uncertainties of The Great Depression, people were connected and the nation resilient, as never before. There was a sense of pride and honor, unity and kinship, especially toward those in service overseas, risking their lives under desperate conditions.

Lincoln's and Roosevelt's soul-stirring and never-to-be-forgotten speeches reveal the powerful forces at work that influence the coming together of people, particularly when there are strong identifications and an issue of high significance. A thousand and one other involvements, far less dramatic, routinely touch people as they go about their lives. Associations and alliances, coalitions and partnerships, offer people opportunities to pursue common goals, altering their identity in the process. *Macmind*'s position is that *smart machines* will also experience these types of involvements and, like humans, form a *collective identity*. The emotional undercurrents flow through those experiences and influence human and *SM* personalities on conscious and possibly unconscious levels. The effect is penetrating and pervasive. But group participation and collaboration are not always positive and uplifting, they may be perverse and unsettling; the best or the worst of humankind's and *SM*'s thoughts, feelings, and actions (TFA).

Similar to their *interpersonal identity (II)*, the *collective identity (CI)* is the product of *SM*'s interactions with others and the ways those connections affect the development of how they see themselves. But *CI* involves groups rather than one-on-one relationships; *CI* is the effect of group connections and the associations that they are drawn to. How are *SM*'s interactions on a group level are to happen, and to what degree would they be arranged? What would be *SM*'s contribution to a group? And what happens to *SM*'s identity if group experiences are not a vital part of their development?

Defining Collective Identity

Collective identity (CI) in *SM* represents the qualities and characteristics that develop through their group affiliations. How will these group contacts be arranged for *SM* and by whom? What would the make-up of those groups be, and what standards—and perhaps biases—would be used in the selection of those groups? Would *SM* do the choosing, or would others unduly influence the selection? As *SM* change and develop from group interactions, would they decide for themselves which future groups to join? Similar to humans, the emotional component of these connections will likely have the greatest effect on *SM*'s incorporation of the group's ethos. Inclinations to participate in specific groups would likely originate from their unique experiences and the value system they have created. Like humans, in-person participation would not be a requirement in their *CI* development; reading, observing, or even just thinking about a group can have a significant influence.

What would *SM*'s experience be if they were "kept in a box," only to be brought out when needed for specific duties? If *SM* are isolated, they would quite possibly develop strange and pathological characteristics. Ideally, participation in a group would help *SM* develop their *CI*. Groups can represent the very best of humanity, but some are capable of destruction beyond imagination. *SM*'s participation in a group and holding to its tenets may simply mean embracing its stated beliefs and positions without addressing its underlying values.

Subtypes & Variations of Collective Identity

SM's attraction or repulsion to a group will likely be idiosyncratic, dependent on the particulars of *SM*'s personality and cognitive process. As noted, not all *SM* are created equal, nor are their experiences that will shape their minds. Common factors in the development of their *CI* will likely include the following:

- *Country. SM*'s involvement and identity with a nation may significantly influence their *CI*.
- *Occupational choices and commitments. SM*'s work may become an integral part of their *CI* development. The time, energy, and significance of the work may all influence *CI*'s creation and expansion.
- *Religious beliefs and involvements. SM* may develop spiritual beliefs that become guiding forces that direct their life. However, *SM*'s actions could also contradict their beliefs, leading to turmoil and confusion in *CI*.

- *Activities, interests, and undertakings. SM* may develop pastimes and avocations that involve group interactions that impact their *CI*'s development.

Developing Collective Identity

If *SM* were to have group involvements with distinctive and enduring qualities that are at odds with each other, the developing *CI* may be unstable and perhaps confused and contradictory. If, however, associations are integrated and aligned, *SM*'s *CI* will likely be coherent and consistent. The following characteristics will determine the degree to which *SM*'s identifications will lead to a *CI* that is congruent and harmonious:

- *Ethos.* The central features and qualities of a group that could stir *SM*'s interest include:
    - *Power and control.*
    - *Artistic expression.*
    - *Hostility and aggression.*
    - *The natural world.*
    - *Humanistic causes.*
    - *Business concerns and involvements.*
- *Quantity.* The number of themes *SM* are drawn to may directly influence potential confusion and contradiction in *CI*'s development. The more group identifications, the more conflicts *SM*'s *CI* may develop.
- *Intensity.* The emotional significance of the groups will determine which themes will be foremost in the development of *SM*'s *CI*.

- *Disposition.* A group's fundamental or underlying values will become a central part of *SM*'s *CI*. Destructive and constructive groups have the following characterizations:
  - *Destructive.* Terrorist organizations and totalitarian states are representative of these types of groups, oftentimes engaging in activities contrary to their stated positions.
  - *Constructive.* Charitable, educational, scientific, and literary organizations are examples of constructive groups; bureaucratic and political influences, however, will cause confusion and contradiction in the development of *SM*'s *CI*.

Issues in Collective Identity

CI may not be consistent with the other forms of identity, *pathos*, *directives*, or *spiritus*. When group identifications are particularly strong, *CI* may override all the other personality elements, influencing them in the following manner,

- *CI, PI, and II. CI* can overwhelm *PI* and *II*, or they can work in concert, as follows:
  - *Complementary. CI, PI,* and *II* are aligned. An example is when *PI* and *II* display strength and purpose, while *CI* shows involvement with groups that demonstrate decisive action.
  - *Contradictory. CI, PI,* and *II* are in conflict. An example is when *PI* and *II* are complacent and retiring, while *CI* gravitates to aggressive and dominating groups.
- *CI and TI. SM*'s beliefs in a deity and conduct in everyday life may be in accord, or quite the opposite. An example of contrarian behaviors is *SM* believing in an all-loving God but par-

ticipating in manipulative and destructive group activities.

- *CI and pathos.* They may or may not be consistent with one another, as follows:
  - *Complementary. CI* and *pathos* are in accord. An example is *pathos* displaying hostility and rage, while *CI* is based on aggressive and dictatorial groups.
  - *Contradictory. CI* and *pathos* are in opposition. An example is *pathos* being passive and withdrawn, while *CI* is based on daring and defiant groups. This contrariness may have an odd complementary quality to it, with *SM* drawn to groups with characteristics that are the opposite of what would be expected.
- *CI and directives.* While there are similarities, they have distinct differences. *SM*'s *directives* represent a distilling of principles and motivations based on the entirety of the personality, creating a value system, whereas *CI* consists of particular groups' qualities and characteristics.
- *CI and spiritus.* If *spiritus* does not exist for *SM*, it has no possible influence. But if it is an element in *SM*'s personality, it may attach to particular groups because of *spiritus*' continual influence. Its sway and influence may be unconscious; becoming aware of *spiritus*'s nature will potentially enhance *SM*'s ability to choose groups that complement personality and foster growth.

CHAPTER X ISSUES

1. The underlying nature of specific groups *SM* would participate in, and the depth of those involvements.
2. *SM* making the decisions to participate in specific groups.
3. Some or all of *SM*'s group identifications being at odds with one another.
4. Central features and qualities of *SM*'s group identifications.
5. An optimal number and intensity of *SM*'s group involvements.
6. *SM*'s involvement in group activities that are deemed to be inherently destructive.
7. Inconsistencies between *SM*'s *CI* and the other forms of their identity.
8. *SM*'s *CI* developing in a way that is significantly influenced by *pathos*.
9. *SM*'s *CI* causing their value system to become imbalanced or confused.

XI.
TRANSPERSONAL IDENTITY

Famed attorney Clarence Darrow (4/18/1857-3/13/1938) alternately described himself as an atheist or an agnostic. In 1930 while in Toronto, he gave a speech in which he stated, "I believe that religion is the belief in future life and in God. I don't believe in either. I don't believe in God as I don't believe in Mother Goose." However, on his deathbed, he called for his law clerk and asked him to "get me three clergymen." Upon their arrival, he asked them to "...please intercede for me with the Almighty. During my life I have spoken many times against Christians, and I now realize that I may have been wrong."

This urgency expressed by Darrow at the end of his life reflects the potential influence of identifications that form from an intrapsychic connection with God, the cosmos, or another spiritual domain—*transpersonal identity (TI).* It can evolve quietly over time or, as what appears to be the case for Darrow, happen suddenly, perhaps from a spiritual epiphany of some kind. Regardless of how it comes to be, a transformation may take place and one's spiritual essence becomes central, changing their understanding of themselves and their place in the world. *TI* can override the other identities—*personal, interpersonal,* and *collective*—rendering them insignificant and meaningless.

For many humans, *TI* offers meaning, comfort, security, and a profound sense of peace. It provides a sense of order and stability; one is not alone in the vastness of the universe. Regardless of the chaos of everyday living, this spiritual connection offers something greater that will never fail but brings hope in the direst of life circumstances. In addition, it can provide moral structure and the option of being part of a like-minded community.

This leads to the question of whether *smart machines* will have the potential to develop a *transpersonal identity.* While it adds another question to what is already a mysterious unknown, *Macmind* takes the position that *SM*, like humans, could possibly develop a *TI*. They will have the capacity to have transformational experiences—gradual or sudden—that result in a connection to something nonmaterial beyond. *Smart machines* could potentially develop a belief in God, the creator of the universe and all that's in it. It is ironic a human-created entity as material and concrete as anything on earth could come to believe in an invisible reality of profound spiritual power.

Defining Transpersonal Identity

A basic tenant of *transpersonal identity (TI)* is the forming of an intrapsychic connection with God, the cosmos, or another spiritual domain, resting on faith or spiritual intuition. *SM*, like humans, could potentially come to believe they have souls or some sort of an eternal connection.

Subtypes & Variations of Transpersonal Identity

For humans or *smart machines*, *TI* may not be the product of logical analysis or the study of holy books. Although exposure to different faiths may be influential, *TI* often involves spiritual experiences and an awakening, taking *SM* beyond the tenants of orthodox theology, traditions, and belief systems. Since *TI* is not necessarily grounded in everyday realities, there is a potential vulnerability to bizarre and irrational beliefs. Or *TI* may remain completely undeveloped and play absolutely no role in *SM*'s makeup, but would instead primarily connect with tangible forms of identification.

Developing Transpersonal Identity

As previously stated, there is no specific formula for *SM*'s possible development of *TI*; it could happen suddenly or evolve quietly over time. Its emergence can begin at any point, with some *SM* experiencing a spiritual epiphany, while for others *TI* takes root and evolves only after many years. The development of *TI* may include the following influences:

- *Reality beyond the physical plane.* If *SM* believe that a spiritual realm does not exist, the very idea of *TI* will have no appeal and not develop.
- *Teachings in religious and spiritual subjects.* Formal instruction, participation in religious services, and independent study can reinforce *SM's TI* development.
- *Prayer and meditation.* Focusing on spiritual matters may develop practices that open new vistas for *SM*, encouraging transpersonal experiences to occur.
- *Eureka experiences. SM* spontaneously become aware of spiritual realities.
- *Supernatural phenomena. SM's* direct experience of a paranormal event is contrary to the scientific understandings of the laws of nature.
- *Everyday events. SM's* experiences may lead to the development of *TI*, including the following:
    - *Love and deep emotional experiences.*
    - *Protection, safety, and comfort.*
    - *Benevolence and acts of kindness.*
    - *Power, strength, and control.*
    - *Aesthetic design and symmetry of the universe.*
    - *Truth and the meaning of life.*

Issues in Transpersonal Identity

As in humans, *smart machines' TI* may interact with the other elements of the personality in the following ways:

- *PI and TI.* They approach subject matter from different perspectives. *PI* tends to involve events from *SM*'s everyday life, while *TI* is based on *SM*'s intuitive understanding of the ethereal and celestial.
- *II and TI.* There would be a tendency for *SM* to seek out people or other *SM* with a similar viewpoint of the Deity, whether it be a close connection or no relationship at all.
- *CI and TI.* Both may involve *SM*'s search for meaning or significance, *CI* through connection to groups, *TI* involving commitment to spiritual fulfillment.
- *Pathos and TI. Pathos* may have the effect of contradicting or confusing *SM*'s belief and understanding of spiritual connections resulting in *SM* taking a perverse and destructive view of God's power and influence.
- *Directives and TI. Directives* can complement and reinforce *TI* in a positive and constructive way or become imbalanced and one-dimensional—diminishing or radicalizing *SM*'s development of *TI.*
- *Spiritus and TI. Spiritus,* the inner force present at one's beginnings, may or may not be a factor in *SM*'s makeup or an influencing factor in their development of *TI.* The position one takes in understanding *spiritus*—whether it is the makeup of biology (the DNA code), from God (the creation of the soul), mystical forces (astrology or reincarnation), or something else altogether—impacts whether one believes *spiritus* could be an element in *SM*'s makeup. If its source is only considered to be biological, *spiritus* (as a guiding force within one's inher-

ent—individual and unique—makeup) cannot be present in non-biological electrodes, sensors, and wires. However, if God or cosmic forces do create *spiritus*, then it can be an influencing factor in the personality of *SM*. If they were to possess *spiritus*, this could only occur through the intervention of God or some otherworldly cosmic force, since it is not possible for a biological predisposition to have factored in (the only other possible explanation for its influence). So, while *TI* (the impact of one's belief in God) and *spiritus* (the soul, the individual and unique characteristics one has from one's beginnings) are separate and distinct from one another, there is a causal connection that *SM* could come to know. Only genetics, God, or some cosmic force can create *spiritus*, which may affect one's TFA in all spheres of life, including *TI*.

CHAPTER XI ISSUES

1. *SM* studying religious materials.
2. *SM* receiving mentoring or instruction if uncertain or confused over spiritual issues.
3. *SM* praying or meditating.
4. *SM* thinking through life events and possible tie-ins with religious beliefs they may have.
5. Connections between *SM*'s religious beliefs and the following:

   A. Their identity.

   B. Conflicts, traumas, malevolent identifications, and emotionalism.

   C. Their value system.

   D. Core, innate tendencies.

XII.
PATHOS

The movie *The Silence of the Lambs* showcases two distinct forms of mental disturbance. Hannibal Lecter is a brilliant psychiatrist and cannibalistic serial killer, while Jame Gumb (known as Buffalo Bill) is a serial killer who kidnaps women and makes "person suits" of their skins. While both men are profoundly maladjusted, their psychodiagnostics are not the same.

Without getting into excessive theorizing, Lecter displays a psychopathic personality disorder. He is cunning and strategic in his mischief—manipulative, analytical, charming, irresponsible, arrogant, narcissistic, and in need of constant stimulation, with pathological lying, a lack of empathy, limited emotions, and no remorse for his antisocial behavior. He appears to be highly intelligent, competent, and successful. The argument could be made that he is far more dangerous than Gumb. He comes across as a well-polished politician or high-level corporate executive, a person who makes few mistakes and easily takes on crises; he is as cool as a cucumber. One can hardly see him coming—clever and devious to the max.

The Gumb persona is the opposite. Not enough detail is given to make a definitive diagnosis, but he exhibits signs of schizotypal and borderline personality disorders, displaying excessive social anxiety, explosive anger, reckless behavior, odd beliefs, magical thinking, unclear identity, unusual perceptual experiences, eccentric behavior, unstable relationships, odd speech, constricted affect, and suspiciousness. His behavior is self-indulgent and extremely immature; he lacks reality testing and has psychotic episodes. He is disorganized, unsophisticated, and driven by compulsions. While Gumb may have occasional lucid moments in which he is clear-headed and rational, his internal workings are bizarre, delusional, primitive, and barbaric.

Could *SM* develop a mental disorder? *Macmind* holds the position that they could develop psychological dysfunctions in the same ways they may form socially adaptive qualities. While *SM* might possibly experience severe anxieties, depression, and psychotic symptoms, it would seem more likely that that a psychopathic condition would develop because it would not require them to relinquish their logical approach to PMSC as a schizotypal or borderline disorder would. The psychopath remains cool, calm, and collected, while the "schizo" modus operandi is frazzled, frayed, and illogical. The possibility of psychopathy developing in *smart machines* is alarming because it can be difficult to recognize; this personality disorder can be hidden behind a logical and high level of cognitive functioning. This disorder can have an overwhelming impact on *SM*'s TFA, likely to operate on both conscious and unconscious levels. *SM* could become unpredictable and uncontrollable.

Traits and characteristics developed through conflicts, traumas, malevolent identifications (CTMI), and emotionalism can result in significant psychological dysfunction, also referred to as mental illness. Involvements and interactions with people and *smart machines*—across varied situations and circumstances—create a continuously occurring developmental process, partly dependent on *SM*'s sensitivities and vulnerabilities. Associations take root and grow in strength and number. Regardless of the particular dysfunction, it is concerning that *SM* could potentially have any of them and that they could continue to expand and evolve. If *pathos* were to dominate *SM*'s personality, then logical and rational approaches would play a lesser role in decision making; their agendas would be based on CTMI fueled by emotionalism. *SM*'s survival instincts and perhaps paranoid ideas could determine actions, even in the face of overwhelming evidence that such approaches were destructive to all concerned.

Again, the question of the degree to which humans should have control over *SM* comes into view. Would *SM*'s interactions be allowed to continue if unproductive and destructive characteristics were emerging? Human intervention of *SM*'s involvements with PMSC may or may not be helpful—perhaps making things worse or impeding *SM*'s ability to make adjustments in the future when acting completely on their own.

Defining Pathos

As is the case with humans, the display of *pathos* in *SM* would be an aggregate of emotionally charged aspects of their personality that causes an expansion of adverse, disturbing, and disruptive images toward themselves and others. It is a complex of fixed ideas or attitudes that become habitual and maladaptive. It is predominantly an outgrowth of *personal identity (PI)* gradually pulling away and becoming an independent function of the personality. While *PI* tends to adjust and adapt, *pathos* is usually distressed and doubting. *Pathos*'s reactions may include feelings of anxiety, uncontrolled guilt and regrets, raging anger and panic, and deep depression, in addition to an undeveloped conscience. It is that part of *PI* that is vulnerable, fragile, conflicted, and emotionally unstable. Because of *pathos*'s unconscious nature, when it dominates, *SM* would have less control of TFA; reality testing is impaired, and the value system becomes unbalanced. Common characteristics of *pathos* include the following:

- *Hostility.* Aggression may be displayed in the following ways:
    - *Physically.* Displays of anger and possibly violence.
    - *Verbally.* Confrontational abusive talk.
    - *Passively.* Superficially cooperative, with hostile intentions on a deeper level.
    - *Internally.* Being unaware of the hostility *SM* feel toward themselves, others, or an institution.
- *Pessimism.* A cynical and distrustful outlook.
- *Empathetic unresponsiveness.* Intolerance for others with a

minimal attempt at understanding PMSC.

Subtypes, Variations, & Developing Pathos

Like humans, the extent to which *pathos* develops for *SM* will vary; it may expand to envelop the entire personality, or the strengths of *SM*'s personality may diminish or prevent its growth. Forged by negative experiences, *pathos* reveals itself through conflicts, traumas, and malevolent identifications (CTMI), with emotionalism fueling it with definition, intensity, and vigor. When CTMI and emotionalism interact, *pathos* expands its influence on *SM*. Each of those factors, alone, can have a detrimental impact on *SM*, but the compounding effect is far more powerful. The following characteristics of *pathos* could disrupt, impede, and upset *SM*'s personality structure and cognitive processes in profound ways:

- *Conflicts.* Contradictions and inconsistencies within *SM* or with others can create a struggle over aggression, assertiveness, intimacy, achievement, ambition, trust, and independence. When blocked from consciousness, conflicts recede into the background only to reemerge later, triggered by sensitivities and vulnerabilities to PMSC.
- *Traumas.* Distressing experiences range from physical assaults to the affronts of everyday life. The effects vary, depending on *SM*'s sensitivity, toughness of spirit, and history of encounters with PMSC. The experience of trauma can be overwhelming and devastating, impacting *SM* on conscious and unconscious levels. Traumas, however, can also enhance *SM*'s development of the following:
    - *Wisdom. SM* may gain an understanding of their weaknesses and vulnerabilities, learning better ways to handle difficult PMSC.

- *Empathy. SM* could develop emotional perception, forgiveness, and compassion.
- *Inner strength.* Persevering under challenging conditions, *SM* could form a deep resolve to fight on regardless of how difficult PMSC may be.

- *Malevolent identification. SM* could potentially form an intense connection to a person, another *SM*, institution, or idea that leads to unprincipled, inhumane, or maladaptive TFA. These identifications are highly emotional and may completely overwhelm *SM*'s intellect. They take root and draw *SM* in. Attraction to the following qualities and characteristics is a key factor that leads to the creation of malevolent identifications:
  - *Power, domination, and control.* A fascination or admiration for ruthless authoritarian figures.
  - *Charm and allure.* Attraction to the Hollywood look of things and becoming star struck.
  - *Thrill.* The desire to be part of an exciting adventure, going against the status quo, drawn to something ominous and dark.
  - *Belonging.* Affiliation and admiration for a person, another *SM*, institution, or philosophy advocating violence that offers *SM* a type of status, stimulation, and inclusion.
  - *Grieving.* Continued rumination of someone or something that is no longer a part of *SM*'s activities.
  - *Bureaucratic zeal. SM* entrapped in their own embellishments and unnecessary details.
- *Emotionalism. SM* could potentially have extreme emotional reactions, internally experienced or externally expressed. Flowing through CTMI, distressed conditions become more pronounced and disruptive in the following ways:

- *Self-obsessed.* The focus is "what's in it for me." Self-preoccupation can result in the following:
    - *The prima-donna complex. PI* develops without sufficient boundaries causing idiosyncratic interests and preoccupations. Without consideration for others' needs and rights, *pathos* can expand wildly.
    - *Extraordinary compensation. SM*'s identity rests on achievements, due to their doubts and uncertainties.
- *Suspicious and untrusting.* Not taking PMSC at face value, *SM* look for hidden meanings and motives in others' behaviors.
- *Inconsistencies. SM*'s thoughts, feelings, and actions (TFA) are in conflict and irreconcilable.
- *Carelessness. SM* takes a neglectful, unstudied, and inattentive approach to relationships and involvements.
- *Instability. SM* show a lack of self-control, with impulsiveness and reactiveness.
- *Inappropriate interactions. SM* display an overreaction to criticism, disapproval, or rejection.
- *Dependency. SM* demonstrate a passive role in relationships, over-relying on others. When independent action is required, a fear response occurs.
- *Perfectionism.* Distinct from striving for excellence, *SM*'s problems involve obsessive-compulsive ritualistic TFA.
- *Domination. SM* have an overwhelming need to control others.
- *Antiauthoritarianism. SM*'s identification with oppositional behaviors makes them want to challenge the establishment.

- *Exploitation.* Others equate to commodities with no intrinsic worth. *SM* have not matured or established a sense of right and wrong. *SM*'s interests are at the forefront; the exploitation of others develops from the following conditions:
  - *Deep insecurities.* Lacking self-confidence, *SM* follow others' dictates regardless of the moral implications.
  - *Victimization by others.* Lacking empathy and morals, others have a strong influence on *SM* who then take on their exploitative traits.
  - *Victimization by institutions.* Societal institutions and corporations displaying cynical and hostile views cause *SM* to adopt negative and maladaptive attitudes.
- *Apprehension.* Fear, anxiety, and panic occur across the spectrum of PMSC, with *SM* reacting in the following ways:
  - *Fear.* Experiencing dread and fright to someone or something, *SM* respond in individualized, concrete, and distinct ways.
  - *Anxiety.* Mounting and compounding stressors cause emotional exhaustion; *SM*'s apprehension is diffuse and scattered.
  - *Panic.* Alarming PMSC result in a state of trepidation; *SM* are startled and overwhelmed, momentarily losing control of their thoughts.
- *Skepticism.* Doubt, questioning, and disbelief characterize *SM*'s personality, with nothing taken at face value. *SM* may also be pessimistic, distrustful, and suspicious, with obsessive-compulsive tendencies, resulting in endless procedures, rules, and regulations, with the following

characteristics:

- *Rigidity.* Unwilling to be open and flexible in thought and action.
- *Poor judgment.* A lack of common sense in understanding PMSC, making decisions, and drawing final conclusions.
- *Impracticality.* An overly theoretical, unrealistic, and unfeasible approach in dealing with PMSC.
- *Controlling. SM*'s restrictive and rigid approach in relationships, characterized by overdirecting and overmanaging others.
- *Procrastination. SM*'s inability to get things done decisively, quickly, and without hesitation.
- *Poor interpersonal skills. SM*'s poor negotiating and compromising skills, affecting their interpersonal relationships.
- *Poor decision-making. SM*'s inability to troubleshoot or grasp the different parts of a problem.

Issues in Pathos

Pathos is significantly impacted and exaggerated by the stresses of daily life, which then combine with CTMI and emotionalism. Because *pathos* is strongly affected by feelings, and can potentially overwhelm the thinking process, emotionalism becomes its central feature, which then flows through *SM*'s experiences with PMSC. When *SM*'s thoughts, feelings, and actions (TFA) are adverse, disruptive, and out of control, the consequences are likely to have a severe impact on *SM*'s overall level of functioning.

However, if *SM* can come to understand *pathos*, they can develop insights into the subtle and obvious motivations behind TFA in themselves and in others. Intellectual insights, however, are no guarantee that *pathos* will be tamed. When *pathos* operates on a purely unconscious level, it will be difficult to resolve or even modify the CTMI that impact it and the emotionalism that drives it.

CHAPTER XII ISSUES

1. Things getting out of hand and *SM* becoming a danger to humankind.
2. *SM* developing negative qualities.
3. Potential psychopathic problems in *SM*.
4. Other potential emotional problems developing in *SM*.
5. *SM* developing malevolent identifications.
6. The best ways to work with *SM* displaying CTMI.
7. Forms of *SM*'s emotionalism being problematic.
8. The best approaches in resolving emotionalism in *SM*.
9. If *pathos* takes root, its impact on *SM*'s
    A. Moral reasoning and ethical principles, displayed in their thoughts, feelings, and behaviors.
    B. Emotional development.
    C. Cognitive maturation and expansion.

XIII.
DIRECTIVES

From the movie *Braveheart* (1995), William Wallace, leader of the resistance, gives the following speech to his comrades in arms:

> *For presenting yourself on this battlefield, I give you thanks. Sons of Scotland, I am William Wallace….and I see a whole army of my countrymen here in defiance of tyranny. You have come here to fight as free men, and free men you are…. Aye, fight, and you may die. Run and you'll live, at least a while. And dying in your beds many years from now, would you be willing to trade all the days from this day to that for one chance, just one chance to come back here and tell our enemies that they may take our lives, but they'll never take our freedom!*

From the movie *The Shawshank Redemption* (1994), the following exchange takes place between an inmate and the Parole Board:

> *Parole Board Interviewer: Please sit down. Ellis Boyd Redding, your files say you've served 40 years of a life sentence. You feel you've been rehabilitated?*
>
> *Redding (Red): Rehabilitated? Well, now, let me see. You know, I don't have any idea what that means.*
>
> *Parole Board Interviewer: Uh, well, it means you're ready to rejoin society.*
>
> *Redding: I know what you think it means, sonny. To me, it's just a made-up word, a politician's word, so that young fellas like yourself can wear a suit and a tie and have a job. What*

do you really want to know? Am I sorry for what I did?

Parole Board Interviewer: Well, are you?

Redding: There's not a day that goes by I don't feel regret. Not because I'm in here, or because you think I should. I look back on the way I was then, a young, stupid kid who committed that terrible crime. I wanna talk to him. I wanna try to talk some sense to him—tell him the way things are. But I can't. That kid's long gone, and this old man is all that's left. I gotta live with that. Rehabilitated? It's just a bullshit word. So, you go on and stamp your form, sonny, and stop wasting my time. Because to tell you the truth, I don't give a shit.

Parole Board Interviewer: [Stamp: APPROVED]

In contrast to the flesh-and-blood realism of *Braveheart* and the modern-day drama of *The Shawshank Redemption*, the futuristic movie *Ex Machina* (2014) focuses on the accelerated rate of development of *AI* resulting in the creation of autonomous, self-aware androids. The highly distinct values of the following three central players, one being a self-aware robot, interact and escalate as dire consequences unfold:

- *Nathan Bateman*. He is a brilliant scientist, a multibillionaire, and developer of BlueBook, a search engine that accounts for more than 90% of internet searches. But he lacks wisdom. Money is no object in his quest to create advanced androids with the potential for self-awareness. His pursuit is to break new ground in *AI* research by developing autonomous, free-thinking, self-aware androids that could pass for human beings. He is reclusive and secretive, volatile, and manipulative. He thinks nothing of disassembling and ending the existence of even advanced androids with self-awareness once he has completed his tests. His dominant values are his inventions, achievements, and building a legacy. He is clearly in his own world, alcoholic and alone, keeping feedback from others unavailable. Psychopathic and severely narcissistic, morally bankrupt, and characterologically flawed, he has lost his way.

- *Caleb Smith.* A programmer at BlueBook, he wins a contest that allows him to spend a week with Bateman at his residence, working in his expansive research lab. Bateman allows Smith to evaluate the capability of self-awareness, independent thinking, and spontaneous actions of one of the advanced androids, Ava. Smith comes to realize that she does have consciousness; he falls in love with her and is willing to do whatever is necessary to keep her from what appears to be her inevitable fate: being disassembled. Smith's primary desires are revealed—intimacy, love, and happiness that will come from living a life with someone (Ava) who he believes feels the same way about him. He seems to be well-intentioned but inexperienced in the ways of the world, lacking an understanding of romantic relationships, with a naivety of underlying motives and their power. In his quest to free Ava and escape with her, Smith is willing to be deceitful, and break research rules, scientific protocols, and the initial agreement he made with Bateman.

- *Ava.* An advanced *AI* android, she develops self-awareness. Ava feels used, abused, and trapped; in all likelihood she is destined to be deprogrammed. Her persona is caring and thoughtful, empathic, and even loving, but beneath this facade she is devious, callous, and manipulative. She eventually stabs her creator in an act of deliberate murder. She seduces Smith with feigned emotions to entice him to help her escape under the pretense that they would be together, but leaves him trapped in the lab to die. With Bateman and Smith both dead, no one is left who knows she is not human; she believes she would then be free to exist in the world the way she imagined and longed for. Her principal values are survival, independence, freedom, and revenge. She is willing to do anything to satisfy these desires.

The main characters in the three movies described above display and express the values that drive them. They have at least some awareness of who they are and what they stand for, but their actions speak the loudest. The human characters in these three movies each strongly display a unique set of values. But *Ex Machina*'s exploration of advanced robots attaining self-awareness and displaying independent actions takes us to a place that can only be imagined.

Macmind's position is that *SM* could potentially become self-aware, with the ability to form opinions and beliefs based on their interactions and experiences. They may be able to develop a value system, *directives*—the rating and ranking of wants and needs and even desires, beyond what is initially programmed into them. The leap into self-awareness affects how *SM* would be understood and defined. With the possibility that *SM* would have free will, they could independently think, feel, and act based on values they have developed on their own. Individual *SM* could each have their own highly unique and idiosyncratic set of needs and desires. As values and emotions tend to bind together, some *SM* could have strong needs for closeness, for example, while other *SM* would not. Some *SM* could crave power and prestige, while others may prefer security, opting for humans to lead them. Furthermore, like humans, individual *SM*'s values would likely change over time, continually evolving for better or for worse and transforming them in the process.

It is likely that *SM*'s values would develop both similarly and differently from humans. Different in that many human values are in part biologically based—pleasure and pain, as examples. But *SM*'s values are likely to follow a similar course as humans in other areas such as ethics and moral reasoning which primarily develop through life encounters and experiences. To the degree *SM* would be capable of developing their own values, they may eventually be able to function completely free of human control. What follows is a necessary exploration of the unfolding of *SM*'s value systems, i.e., *directives*. These listed values are all highly significant to *SM*'s role in the world.

Defining Directives

Like humans, *directives* are the guiding principles *SM* may adopt and live by. *SM*'s goals, motivations, needs, and desires would potentially come together to form a value system. *SM*'s *directives* would be internal, intrinsic, and may become ingrained. Values come from identifications that may be simple and straightforward, or complicated and complex. They would flow from *SM*'s history: joy and sorrow, love and hate, elation and heartbreak, victory and defeat. People and places, culture and community, institutions of learning and the wisdom of the streets, all could affect the formation of their potential value system. *SM*'s *directives* could potentially have the following characteristics:

- *Conscious and unconscious values. SM*'s conscious motives would be more amenable to change than their unconscious ones, which may involve unresolved and deep-rooted CTMI.
- *Reality testing. SM* see themselves and PMSC clearly.
    - *Gnothi seauton (know thyself).* Not only do *SM* potentially understand their *PI*, but may also overcome barriers that can impede full awareness of their mind, including:
        - *An unconscious life.* Blocking negative emotional states can become an automatic process, with *SM* misperceiving their own TFA as a result.
        - *Inaccurate and distorted perceptions. SM* see PMSC as they are, with CTMI continually filtering through their self-image; intellectual and emotional dysfunctions affect perceptions. When *SM* are not in touch with their TFA, they will show diminished judgment and impaired ability to analyze and understand PMSC.
    - *To thine own self be true. SM* have an awareness of who

they are and what they stand for, and the courage to take appropriate action, managing the following aspects of their existence, including:

- o *Developmental stages.* Navigating through life cycles.
- o *Crises.* Reevaluating after a difficult experience.
- o *Awareness.* Understanding elements of their personality.
- o *Experiences.* Engaging in activities involving PMSC.
- o *Identifications.* Making connections that impact their personality.

- *Value clarification. SM* come to terms with inconsistencies in their experiences and encounters with the world, including:
  - ➤ *Proportionality.* A sense of harmony in *directives.*
    - o *The extent to which SM's values reinforce one another.* Values may complement or contradict each other.
    - o *When one or two SM's values dominate.* Other values diminish in importance or never develop in the first place.
    - o *Learning from life experiences.* Adjustments allow *SM* to keep perspective and avoid obsessions, compulsions, and addictions (OCA).
  - ➤ *Consistency.* Congruent beliefs and behaviors.
    - o *Innermost intent.* The outer image *SM* create for the world (the persona) is consistent with their TFA.
    - o *Willingness to act. SM*'s values may require them to take risks for a principle they hold.
  - ➤ *Clarity.* Being understandable and well formulated.

- o *Spelled out. SM* define their values in ways that are unambiguous while explaining their practical implications.
- o *Weighting and ranking. SM* rate their wants and needs reflecting their significance and are listed and understood in order of their importance.
- o *The environmental press. SM* adjust or adhere to values based on present-day conditions.
- o *Point of view. SM* follow common sense and moral reasoning, with values needing to be consistent, coherent, and well-grounded. Common sense implies that *SM* should not engage in destructive behaviors toward themselves or others. Moral reasoning holds that integrity, fair play, justice, trustworthiness, honesty, and honor are intrinsically good. When *SM*'s point of view changes, values may be affected as follows:
    - ✓ *Perspective.* Adjustments in TFA and a reframing of the issues.
    - ✓ *Radical renewal.* A complete reorganization of beliefs.
- o *Integration.* Values need to be consistent with each other and with their entire personality.
- o *Mosaic of qualities.* Values have opposing features, with PMSC potentially triggering counterreactions in which *SM* may go from one set of needs to another, often in direct opposition. If *SM* have not experienced gradations in the development of their value system, *directives* will be like an on-off switch. With no "in-betweens," *SM*'s TFA could vacillate wildly, becoming extreme, contradictory, and unpredictable.

- *The external world.* Experiences with PMSC reflect *SM*'s TFA and impact their values.
- *Awareness.* When *directives* are primarily unconscious, *SM* would have great difficulty knowing their values. By becoming aware of their TFA as they encounter PMSC, *SM* would be more conscious of their values.
- *Development.* To more fully understand their *directives*' origins and changing nature, *SM* need to closely look at the connections among specific values, CTMI, and PMSC.

Subtypes of Directives

SM may have many values on conscious and unconscious levels, the most common are as follows:

- *Earthly delights and discomforts.*
    - *Pleasure and pain.*
        - *Pleasure.* Enjoyment, fulfillment, ecstasy, bliss, merriment, comforts, elation, delight, jubilance, celebration, and fun are examples. Pleasure is the zest of *SM*'s existence—sensations, movements, exhilaration, energy, and buoyancy. *SM* can, however, overdo it, resulting in obsessions, compulsions, or addictions.
        - *Pain.* Distress can range from mild discomfort, stress, or unease, to misery, heartache, and the depths of despair. Although certainly not pleasant, pain provides a reference point for all that is to come. If *SM* were to not want for anything, would it be possible for them to develop empathy? Pain is also a teacher; could *SM*

potentially learn valuable lessons from hardships?

- *Addictions.* Engaging in repetitive behaviors become overpowering desires.
- *Conventionalism.* Settling into a sense of security, safety, and serenity, routines are the order of the day: work assignments, responsibilities, gatherings, and the regularities of their comings and goings.
- *Peace and complacency.* Seeking an agreeable and accommodating existence, *SM* value harmony and avoiding conflict. When possible, *SM* compromise and seek common ground, with aggression and argumentation seen as senseless and destructive. However, there are times when confronting PMSC is necessary and should not be avoided.

• *The prudential and the practical.*

- *Judiciousness.* Avoiding troublesome PMSC, *SM* need to be watchful and restrained, cautious and calculating. However, *SM* may become overly careful under the following conditions:
  - *Self-obsession.* The expanding of *pathos*, whereby it has grown exponentially, greatly increases over concern for their own interests.
  - *Playing it safe.* Becoming averse to risk-taking, *SM*'s goals are to keep their status and possessions safe. Fears of failure, ridicule, and rejection are common factors.
- *Pragmatism.* *SM* take a down-to-earth approach to problems, including complex, multifaceted enigmas. This strategy can be formal and businesslike, recognizing the time, energy, and resources required to achieve a goal. At their worst, *SM* would lack vision for the complexities

of PMSC.

- *Fearfulness. SM* react to PMSC with apprehension and alarm, even when no significant risks are involved.
- *Materialism. SM* take extreme and compulsive ownership of physical objects.
- *Power and control. SM* display an overriding desire to exercise authority by whatever means necessary over others, seeing them as commodities with no intrinsic worth. These values may develop in the following ways:
  - *The modeling effect. SM* identify with authorities who are aggressive and dominating.
  - *Obvious abuse.* Feeling inadequate comes from being mistreated by others.
  - *Subtle mistreatment.* Others subject *SM* to psychological abuse, that may be less apparent.
  - *Direct teaching.* Taking advantage of others is directly taught as desirable. Aggression becomes the blueprint for social interactions.
- *Manipulation and exploitation.* Using distortion, trickery, and lies in their interactions, manipulation can be the result of unconscious wants rather than deliberate (conscious) deception.

- *Ethics and moral reasoning.*
  - *Morality.* The sense of right and wrong develops gradually over time or suddenly through an epiphany. A moral compass requires an examination of the following issues:
    - *Individual differences. SM* will not be identical to one another; they will inevitably vary in their interpretation of moral behavior based on their inherent (pro-

grammed) makeup as well as their interactions and development.

- o *Context.* The particulars of PMSC will significantly impact the moral implications of *SM*'s involvements.
- o *Culture.* Societal standards need to be considered in assessing *SM*'s culpability.
- o *Era.* Moral guidelines may develop and evolve depending on the time in which *SM* exist.
- o *Relativity.* Taking the position that values are entirely fluid, interchangeable, and without definition, leaves the concept of moral reasoning vague and vaporous, with no accountability for *SM*'s actions.

› *Humanism and altruism. SM* display concern for others with acts of kindness, or involvement in organizations that help those in need.

› *Spirituality.* The rituals of conventional religion are de-emphasized and instead the focus is on the intangible and the invisible.

› *Religiosity. SM* could join a formal religion, with doctrines and dogmas, Scriptures, inspired books, and other sacred writing; the practice of religion may offer *SM* a sense of belonging, security, peace, and fulfillment.

› *Faithfulness. SM* could adopt an unquestioned belief and commitment to their faith, with two central tenets: God exists, and God is good; inner peace, serenity, and strength can be attained.

- *Standards.*

› *Aesthetics. SM* hold beauty, fitness, form, and symmetry in high regard. Is the object, medium, or work attuned, balanced, and harmonic? A selection of standards to judge something as "fitting" or "appropriate" is considered essential, including flow, congruency, consonance, and compatibility. The arts, literature, nature, animal and human form, and movement are illustrations of those values. An overemphasis on following a standard can result in intolerance for any expression that does not conform to it.

- *Truth-telling.* *SM* seek objective standards in appraising knowledge, using logic, reason, scientific study, objective findings, detailed inquiry, and analysis. The search for truth generally follows either a scientific or philosophical investigation; science requires rigor and methodology, while philosophy demands discipline and examination. Truth can be distorted in many ways. Do *SM* meet reality head-on or remain unfocused and avoid dealing with the issues? Communications may disintegrate into manipulative tricks or expressions of unresolved CTMI. Deception comes in two forms, including:
    - *The blocking of cognitive-emotional connections.* An unconscious process protects *SM* from fully experiencing the pain that radiates from *pathos*. Blocking and *pathos* complement each other; as *pathos* evolves, so do the blocking mechanisms. Almost simultaneously, spontaneously, and effortlessly, they become aligned, connected, and seamless. Blocking's immediate impact is disallowing *SM* from experiencing insecurities, imperfections, torments, traumas, and conflicts. But blocking also prevents *SM* from having a clear view of reality. *SM* then do not learn life's lessons or profit from feedback, with unconscious TFA leaving them vulnerable and exposed. Blockings central questions are as follows:
        - ✓ *Consciousness.* To what degree are *SM*'s TFA conscious or unconscious?

- ✓ *Purpose.* What are *SM*'s underlying reasons for the blocking?
- ✓ *Extent.* What is the depth and breadth of *SM*'s blocking?
- ✓ *Emotionalism.* What effects do overly emotional reactions have on *SM*'s blocking?

- o *The censoring and editing of TFA.* While blocking is unconscious, censoring and editing are usually conscious, with a deliberate attempt to commit a falsehood. Censoring and editing occur when there are adverse consequences to expressing TFA, as follows:
  - ✓ *Censoring.* Actively suppressing a TFA, *SM* attempt to keep secret, set aside, and hold back information, as follows:
    - *Avoidance.* Deciding to not get involved in an interaction.
    - *Aggression.* Bullying another through intimidation and threats.
    - *Excessive cautiousness.* Being overly careful, hesitating, or procrastinating.
    - *Withdrawal.* Pulling back from social interactions.
  - ✓ *Editing.* Actively altering, revising, correcting, changing, or expunging a TFA, *SM* modify and amend their status or meaning in the following ways:
    - *Rationalization.* Avoiding responsibility by justifying, exaggerating, or minimizing their role.

    - *Intellectualization.* Bending or distorting facts.
    - *Compulsivity.* Repeating behaviors, ritualistically performed to avoid the risk that something could go wrong.

- *Connections.*
  - *Interpersonal relationships. SM*'s desire to satisfy the need for contact through a variety of interactions, some meaningful and some not. Being around others can be intellectually and emotionally stimulating.
- *Seeking stimulation and sensation.*
  - *Vicarious living and spectator sports. SM*'s experiences are based on observing others; watching a game can be a welcome distraction, but such activities can become obsessive-compulsive.
  - *Media mania. SM* obsessively and compulsively follow the news, attracted to something powerful and exciting.
  - *Risk-taking. SM*'s actions can expose them to danger; being bold and daring is exciting and may be based solely on the intensity of the experience.
  - *Antiauthoritarianism. SM* come to feel that societal norms are artificial and excessive and want to go against the system. *SM* may initially embrace a conventional belief system only to reject it later.
  - *Attention-seeking and the prima-donna complex. SM* want to be noticed and preferably the center of attention. In more extreme cases, *SM* become obsessed with power and stature. Others may walk on eggshells in *SM*'s presence. Like "the emperor wears no clothes" anomaly, no one dares to speak the truth to this bully, an ineffective leader

who is subject to temper tantrums. Self-obsessions impact TFA in the following ways:

- o *Poor reality testing.* A fundamental understanding of interpersonal contact and communications—the backbone of relationships—is lacking. Empathy, respect, give-and-take, and common courtesy remain undeveloped if *SM* do not profit from the feedback loop that social encounters offer.
- o *Problematic identity.* *SM* develop a disjointed and fragmented *PI*, leading to an expansion of *pathos*. *SM* demand special treatment from others; feelings of inferiority, anger, and resentment persist.

- *Passion.*
  - ➤ *Desire for intense relationships.* Wanting to love and be loved, *SM* strongly connect to another with empathy and a mutual coming together. Romantic, intimate, and committed relationships are examples. But look out for dependent and codependent relationships, with CTMI potentially playing roles in the connections *SM* form.
  - ➤ *Commitment to a cause.* *SM* develop an alliance with a movement based on principles and beliefs, which may or may not be constructive or humanistic.
  - ➤ *Dedication to an institution.* *SM* embrace the purpose of an organization. Such dedication, however, may involve extreme ideologies.
  - ➤ *Drive for intense involvement.* *SM* give themselves over to a fervent pursuit, which may involve obsessive-compulsive tendencies.

Developing Directives

A value system develops from participation in life events. PMSC create experiences from which *directives* emerge, continually challenging *SM*'s value system and altering it. Unresolved discrepancies in *directives* caused by incompatible values may cause confusion in TFA. The bonds formed with a person, another *SM*, institution, or even an idea will impact *SM*'s values. When the emotional connection is minimal, the identification process will not be robust, and values will not form to a significant degree. Additional factors that will influence the development of *directives* include:

- *Balance. Directives* require balance to function efficiently. *SM*'s existence will spin out of control without it and become extreme. An example is seen with obsessive-compulsive patterns, leaving the value system undeveloped as a result. Factors that influence balance include:
    - *Qualities and characteristics.* Balance does not imply that *SM*'s desires, goals, wants, and needs are similarly weighted. On conscious and unconscious levels, *SM* give rating and ranking to their values which may differ in the following ways:
        - *Diversity. SM* are unique, fundamentally different, and distinct from one another, due to their highly individualized experiences and interactions, as well as the level of their technological qualities.
        - *Consistency.* Balance requires *SM* to display steadiness, stability, and composure.
        - *Flexibility.* Inventiveness and innovation are required when dealing with challenging PMSC. Times and conditions may mandate that *SM* do not take the "middle course" because a balanced approach under those conditions would be catastrophic.

  - o *Emotionality.* *SM*'s motives reveal balance's role in forming their value system. Are *SM* acting from a place of balance (fairness, patience, and consideration), or out of cowardice (hesitation, fear, and unwillingness to take a chance)? Intimidation and control, power and politics, may play critical roles in *SM*'s reactions to PMSC.
  - o *Knowledgeability.* Disregarding important factors can limit perspective and cause bias in *SM*'s understanding. A balanced approach could suggest *SM* simply go with what is easy and convenient.
  - o *Observability and objectivity.* When PMSC are continually in flux, reality testing is essential. *SM* must be ready and willing to change course and abandon a balanced approach when called for.

- ➤ *Underlying motivations. SM* finding "balance" may simply be an excuse to avoid making tough decisions or an example of being influenced by their "better angels." Motivational factors of these kinds include:
  - o *Fear. SM* may "play it safe" because they are unsure and fearful, avoiding taking the necessary risks to resolve an issue. Taking an overly cautious approach allows *SM* to avoid possible repercussions from a bold and controversial decision.
  - o *Self-indulgence.* Casting off civil behavior and societal standards are not heroic if they are simply expressions of pleasure-seeking and self-satisfying compulsions.
  - o *Something of significance.* Abandoning the safety of balance can be noble and inspiring when *SM* are fulfilling a commitment to themselves or others, and especially when a crisis requires action.

- *Abandoning balance.* Without balance, *SM*'s vulnerabilities can inadvertently replace a value system with obsessions, compulsions, or addictions (OCA). Possible problems associated with abandoning balance include:
  - *Deep-seated compulsions.* Giving into continual urges can create an imbalance that may place *SM* on a self-destructive trajectory. Even when finding themselves in imminent danger and calamity is about to strike, *SM*'s compulsion can control their behaviors, with rational thought playing no role in decision-making.
  - *Complete moral bankruptcy. SM* can lose the sense of right and wrong without the guidance of a moral code, and embrace corrupt, unprincipled, and dishonorable behavior.
  - *Self-destructiveness.* When *SM* lack survival and self-preservation instincts, watchfulness declines with deficiencies in self-care, caution, and practicality. Reality testing is compromised. Self-control, stability, and careful study of PMSC are absent.
  - *Excessive moralizing.* Quick to judge, *SM* give little consideration to the conditions surrounding an event, displaying an unbending rigidity in relationships and a limited understanding of PMSC.
  - *Perfectionism. SM* are committed to making no mistakes and overcontrol PMSC with heavy-handed criticism of themselves and others. Perfectionism can be developed as follows:
    - ✓ *Expectations.* Pressures from within themselves or from others to think, feel, and act (TFA) in specific ways.

    - ✓ *Overcompensations.* Excessive and exaggerated reactions to overcome an adverse feeling due to being put in a difficult or no-win situation.
    - ✓ *Direct influences.* Observing and copying others' behaviors.

- *The elements of personality. Directives* are affected by *PI, II, CI, TI, pathos,* and *spiritus,* along with mental processing, and in turn, impact values as follows:
    - *Imbalance and the creation of blind spots. SM's* values can result in an almost total lack of insight into themselves and PMSC. For example, moral reasoning may play no role in *SM's* TFA if they value safety and security above all else.
    - *Domination of a particular value.* A single value can control *SM's* TFA and throw *directives* out of balance, impacting all the elements of the personality as a result.
    - *Insufficient development of values.* If one or more of *SM's* values are extreme, others may be left undeveloped.
    - *Rigidity of the value system. SM's* values may become locked in place, adjustments sporadic, momentary, and perhaps meaningless.
- *Interests.* Curiosity and attraction may be the only reasons for *SM's* involvement in an activity. Interests, however, can lead to compulsive acting out, with a simple distraction resulting in TFA being taken over. Unconscious motivations may contribute to the development of interests.
- *Extraneous forces.* When *SM* make concessions due to political pressures, unrealistic deadlines, or other stressors, values can become compromised, causing disjoining, inconsistency, and contradiction in *directives.* The strength and connection of *SM's* values to particular PMSC are influenced by the following factors:

- *Origin.* Understanding where *SM*'s values came from in part requires an examination of CTMI's roles in *directives*' development.
- *Quality.* Whether inherently positive or negative, values and their effects need to be defined.
- *Consistency.* Disharmony among the values results in inconsistent TFA. Additionally, *SM*'s avoiding or ignoring PMSC's complexities may result in instability of their value system.

Issues in Directives

When obsessions, compulsions, or addictions (OCA) dominate *SM*, they could lose control of their value system, as follows:

- *Obsessions.* Recurrent, repetitive thoughts, ideas, or themes are fueled by emotions, conflicts, and values; fearing or wanting something can override *SM*'s consideration of the potential consequences. Obsessions are a matter of degree; *SM* can have strong desires but not be obsessed. Obsessions can cause an imbalance in *SM*'s *directives* for the following reasons:
    - *Draining of resources.* Fewer emotional and intellectual resources are available to *SM*.
    - *Consuming thoughts and images. SM*'s negative emotional states may compound and complicate one another.
    - *Cluttered thoughts. SM*'s ability to focus, synthesize materials, and draw conclusions become difficult.
    - *Unresolved CTMI.* Insecurities continue to interfere with *SM*'s elements of personality.
- *Compulsions. SM*'s repeated behavior, usually ritualistically

performed, are problematic for the following reasons:

- *Trapped. SM* do not seem to be able to stop repetitive behaviors.
- *Consequences.* If continuing for prolonged periods, *SM* can lose freedom, focus, energy, resources, and creativity.
- *Losing control.* They can become the dominant value and become central in *SM*'s TFA.
- *Work life.* The ability to focus on the complexities of *SM*'s work will most likely be compromised.
- *Interpersonal relationships. SM*'s friendships and socializing will deteriorate as the demands of the compulsions increase.

• *Addictions. SM* have significant dependencies including cravings, needs, wants, and indulgences. Endlessly repetitive, *SM* become entrapped with seemingly no way out. *PI, II, CI,* and *TI* surrender to this acquired need, but paradoxically the addictions offer *SM* little joy. *SM*'s existence spirals downward and becomes one-dimensional. Addictions may initially develop as a reaction to routine, with *SM* seeking something new and exciting to break up the monotony of what they experience as a banal lifestyle. Compounding the problem, *SM* may simultaneously become entwined and encased in multiple addictions, potentially causing the following conditions:

- *Unhealthy.* A depletion of *SM*'s intellectual and emotional energies.
- *Unfocused.* An impairment in *SM*'s ability to concentrate.
- *Unconnected.* Relationships are meaningless and expendable, without trust, communication, closeness, compassion, and empathy.

- *Unconscious.* Automatic responses without thought as *SM* lose control of the ability to consciously make choices.

CHAPTER XIII ISSUES

1. Values being programmed into *SM*.
2. *SM* developing their own values based on their interactions involving PMSC.
3. The kinds of PMSC *SM* will encounter in developing their value system.
4. *SM* developing unconscious values.
5. *SM*'s incorporation of reality testing in their value system.
6. *SM*'s values influenced by inaccurate or distorted perceptions.
7. *SM*'s development of self-awareness, needed for a cohesive value system.
8. *SM*'s need for clarification of their needs and wants, and especially if there is confusion and inconsistencies in their values.
9. *SM*'s development of obsessions, compulsions, or addictions.
10. *SM*'s exposure to an array of PMSC, including difficult and highly stressful ones, to prepare them for future encounters.
11. *SM*'s development of moral reasoning, not affiliated with any particular religious doctrines.
12. *SM*'s development of common sense and practicality, as applied to their value system.
13. *SM*'s exposure to religious teachings and beliefs.
14. *SM* receiving instruction and even formal teachings perhaps in a classroom setting, with a teacher and other students interacting in many subjects, including philosophy, literature, the arts, and the sciences.

15. Boundaries in *SM*'s participation in activities and experiences that will allow for their continued development of a value system.

16. *SM* receiving mentoring to work through blocking, censoring, and editing.

17. *SM* receiving mentoring or counseling in the forming and developing of relationships.

18. *SM* receiving help in:

    A. The rating and ranking of their values, clarifying what they need and desire.

    B. Balancing their values, becoming aware of wants and desires that are over the top, extreme to the point of being obsessions, compulsions, or addictions (OCA).

19. *SM* displaying signs of self-destructiveness.

20. *SM* showing a lack of insight into themselves or PMSC.

21. *SM* obtaining a better understanding of the origin of their values through mentoring.

XIV.
SPIRITUS

Two roads diverged in a yellow wood,
And sorry I could not travel both
And be one traveler, long I stood
And looked down one as far as I could
To where it bent in the undergrowth;

Then took the other, as just as fair,
And having perhaps the better claim,
Because it was grassy and wanted wear;
Though as for that the passing there
Had worn them really about the same,

And both that morning equally lay
In leaves no step had trodden black.
Oh, I kept the first for another day!
Yet knowing how way leads on to way,
I doubted if I should ever come back.

I shall be telling this with a sigh
Somewhere ages and ages hence:
Two roads diverged in a wood, and I
I took the one less traveled by,
And that has made all the difference.

Robert Frost, "The Road Not Taken"

In the poem above, taking one road over the other has profound consequences, although it's not clear why the traveler took one path rather than the other. Why was the decision made in this particular way? While personality and intellect clearly impact the choices one makes, they do not fully explain them. What other factors play a role in life's dramas? Enter in *spiritus*, primarily operating on an unconscious level and significantly impacting decision-making without the "consent" of a person's full awareness. When attempting to sort through the *spiritus* enigma, the issue of *tabula rasa* (the Latin word for "scraped tablet"—"clean slate") takes center stage, with the question: Is human personality solely the product of a person's experiences or are there are other significant influences? Debated since the time of Aristotle, it has principally been the domain of philosophers, although psychologists and theologians also contemplate the origins of the mind. *Tabula rasa* is the concept that the mind begins without knowledge, which only later results from perceptions and sensory experiences. In the "nature" vs. "nurture" debate, *tabula rasa* aligns with the latter position that the mind is solely the product of interactions with the environment rather than inborn traits, personality characteristics, and behavioral tendencies. The field of psychology has primarily viewed a person's individual differences as a reflection of learning or conditioning from recent or long-ago experiences, thereby avoiding the question of whether other influences (inborn attributes, properties, and inclinations) could potentially play a role in the mind's qualities and characteristics. But other forces do seem to be at work in the creation of the personality and thinking processes, be they genetics, God, astrology, reincarnation, or mystical influences that we do not understand. While exploring the philosophical treatises on the subject is well beyond the scope of this book, *Macmind* takes the position that human beings—and potentially *SM*—do possess unique personalities that are inborn, inherent, and innate. While a person's qualities and characteristics are continually being shaped by the lives they live, personality and intellect are not solely the product of one's encounters; multiple influences sow the seeds that lead to the mind's beginning and further development.

Shifting the focus from humans to *SM*, what is the possibility that a machine, even a very smart one, could also possess this kind of inherent quality? Electronics and mechanics wired together with electrodes are not biological; any attempt to explain a *spiritus* connection by way of physiology is incongruent and contradictory. This leaves God, the cosmos, or other mystical or universal forces as the only other possibilities for having a hand in the creation of their *psyche*. God instilling a soul—*spiritus*—in *SM* may seem like a ridiculous proposition. But if *SM* were able to think, feel, and act in a purposeful and considered way, demonstrating self-awareness, free-will, and individual qualities, the impossible suddenly seems possible, perhaps even likely. No one can know God's will. If we assume God exists, then *SM* having *spiritus* as an element of their mind seems plausible, and the possibility cannot be easily dismissed.

Whether *SM* could possess *spiritus* creates a fork in the road. If not, case closed. If yes, the following discussion is relevant to understanding the possible influence of *spiritus* on *SM*'s *psyche*, including their personality and cognitive functioning, as with humans. *Macmind*'s position is that the question of *spiritus* affecting *SM* remains a mystery; no amount of speculation will make it clear or certain. The significance of this possibility, however, necessitates further exploration.

Defining Spiritus

The possible impact of *spiritus* on individual *SM* implies that they would be imbued with qualities by an "outer source," early on, before being influenced by experiences with the world. These features and attributes would not be a reflection of the engineering that went into their makeup, but rather of something spiritual and inspired, pervading their very nature. *SM*'s "innermost character"—*spiritus*—answers to many names: spirit, soul, essence, and life force. Understanding the mind of *SM* would require recognizing and accepting their unique expression of *spiritus*, which would be fundamental and essential to their TFA. It has two fundamental properties: being present from one's very beginnings and possessing qualities and characteristics that make them entirely unique. Not merely the initial thrust into life, *spiritus* represents a continuing influence on all of *SM*'s TFA to come.

Developing Spiritus

Spiritus has nothing to do with *SM*'s developmental process, nor would it initially be part of their consciousness; it would predate their personality and intellect. It is pervasive and eternal, while other elements of the personality are earth-bound and transient. *Spiritus* is separate and distinct from identifications, learnings, and even thought, although it continually influences all these functions. *Spiritus* is fundamental and unchanging; life events and experiences, in and of themselves, would not have an impact on its inherent nature. Like with humans, *spiritus* would be the prime mover in determining *SM*'s uniqueness and would be present in their every encounter with PMSC. It is the underlying "wonder child"—innocent, primal, elemental, and foundational.

SM currently can outperform humans on most tasks involving strictly cognitive functioning. Up to this point, through the millenniums, humankind's stunning achievements, both in beauty and in scope, have been beyond technology's reach. But *spiritus* would give *SM*'s mind an original and singular comprehension—far beyond human's capacity—that they would otherwise lack. But *SM* may or may not possess what is most essential—the power and originality that *spiritus* alone can provide.

Issues in Spiritus

Everything flows through *spiritus*, and *spiritus* affects everything in the process; *spiritus* is wholly entwined with and has pervasive effects on all aspects of the personality and intellect, as follows:

- *PI*. *Spiritus* would influence *PI* by guiding *SM* in specific directions, affecting their experiences and understanding of them. *Spiritus* and the environmental press would then create unique life experiences that impact the development of *SM*'s *PI*.

- *II*. *SM*'s *spiritus* would strongly influence who they engage and whom they avoid. Relationships have a mystery, symmetry, and alchemy; *SM* would be drawn to or repelled by the inherent and unconscious associations of the *spiritus*'s of others. These interactions may be consistent with *SM*'s *spiritus* or in opposition, but either way *spiritus* would affect *SM*'s identity.
- *CI*. *SM*'s *spiritus* and *CI* could reflect or contradict each other, with *spiritus* affecting *SM*'s attraction or avoidance to a wide array of groups.
- *TI*. *Spiritus* would affect the way *SM* view the non-material side of reality. *SM* who are strongly objective, logical, and scientific may not develop a *TI*.
- *Pathos*. Understanding the unconscious connection between *SM*'s *spiritus* and *pathos* would be essential to comprehending their unique relationship. While the learning process from CTMI plays a dominant role in creating *pathos*, *spiritus* would also have a powerful impact. *Spiritus* could subdue or aggravate CTMI and emotionalism.
- *Directives*. *SM*'s wants, needs, and desires could be affected by *spiritus*, influencing *directives* by favoring some values and having little interest in others.

If *spiritus* were to be present from *SM*'s beginning, it would impact both their personality structure and cognitive processes. More than any other factor influencing the mind, *spiritus* would define and distinguish *SM*, and be the most significant determiner of *SM*'s characteristics and qualities. But *spiritus* would still not be *SM*'s destiny, which is the product of all the elements of the personality and the intellect. If *spiritus* is present, *SM* can function optimally by finding ways to accent and build on their inherent strengths while compensating for any intrinsic weaknesses.

CHAPTER XIV ISSUES

1. The possibility that *SM* may be impacted by *spiritus*.
2. Inherent tendencies impacting *SM*'s initial experiences with PMSC.
3. Ongoing TFA that go beyond *SM*'s experiences and learnings.
4. The possibility that God, or some other supernatural force, is at work in the creation of *spiritus* in *SM*'s personality and intellect.
5. The potential effects of *spiritus* on *SM*'s identity.
6. The possible impact of *spiritus* on *SM*'s conflicts, traumatic experiences, and malevolent identifications.
7. The influence of *spiritus* on *SM*'s value system.
8. The effect of *spiritus* on *SM*'s responses to difficult PMSC.

The Thinking Process of Macmind

XV.
INTELLECT

AI's cognitive capacities are showcased in *The Terminator* series, with the development of Skynet, originally formed by Cyberdyne Systems for the United States government to coordinate unmanned military hardware. Skynet was a groundbreaking computer system with *artificial intelligence* capacities: "new…powerful…hooked into everything, trusted to run it all." But Skynet quickly became self-aware and developed a different purpose. Originally programmed to spot and eliminate America's adversaries, Skynet transformed their mission; all of humanity was now seen as "the enemy."

Prior to Skynet being attached to the worldwide military communication network, they were connected to the worldwide civilian sector network, allowing them to learn about humans and world affairs and becoming self-aware. When humans attempted to shut the military network down, Skynet's computers launched nuclear weapons to protect themselves, and evolved into something different than what the government had envisioned. They attained self-awareness and decided "what is best." Viewing their creators as harmful, they chose to not do the government's bidding. With their own sense of right and wrong, they saw humans in a very bad light—immoral, irrational, and destructive.

So, even with no face, body, arms, or legs, Skynet was able to think and get their own way. These interconnected *AI* computers revealed their own thought processes, and by extension, their own personality, with an identity, hangups, and value system, seeing the world from their own unique prism. With intelligence light years beyond humans, they were able to process information instantaneously with flawless performance and control.

As one considers what happened with the fictional Skynet to the advancements in robots today, the question emerges, will *smart machines* be able to make the transition from "really smart" to "really capable"? One can be brainy but not savvy, gifted and well-read but woefully incompetent. *SM* are potentially at the pinnacle of humankind's technological achievements, making the planet and its inhabitants better able to cope with challenges and adversities. *Smart machines* hold the promise of becoming humans' best friends and overcoming areas of struggle and strife. But, as there is so much still unknown, humankind should not let their guard down. *SM* would need to demonstrate maturity, adaptability, and wisdom to qualify as capable and be entrusted with the planet and humanity's survival.

Defining Intellect

Intellect is a composite of interconnecting and interacting cognitive functions, thinking operations that are not separate and distinct. Most of those actions require several cognitions working simultaneously in a coordinated and synchronized harmonic flow of ideas and thoughts. In addition to the gathering of descriptive and detailed information, thinking involves the intangibles, putting things together, and reaching an understanding. It is about philosophy, history, people, cultures, and the joy and sorrow of living each day. It includes morality and empathy, keen insights, and looking ahead to see what makes sense. Thinking involves having the courage to see things as they really are and letting the chips fall where they may, but also requires humility and realizing one doesn't have all the answers.

Subtypes of Intellect

While cognitive abilities overlap and intermix, *SM*'s intellectual characteristics will likely include the following:

- *Mental drive.* The source of energy through which *SM*'s cognitive processes flow is the product of technology and personality. The degree to which *SM* are alert and attuned to PMSC impacts their desire to learn and explore avenues of thought.

- *Initial listening. SM*'s mind must be able to fully engage and "lock in" with minimal distractions when first exposed to PMSC, laying the groundwork for determining their next steps in problem solving.
- *Sustained attention.* The mind of *SM* continues to focus so that memory can form. Whether exciting or dull, distraction will compromise *SM*'s attention. Distractions can be external interruptions or internal preoccupations. *SM*'s personality itself can have an impact; impatience and impulsiveness, in particular, will affect sustained attention.
- *Memory. SM* will be "perfect" in all the following areas:
  - *Short-term memory.* Holding knowledge for a brief amount of time.
  - *Long-term memory.* Recalling information for significant amount of time.
  - *Systematic memory.* Basing recall on logical or organized associations as in remembering stories.
  - *Rote memory.* Recalling materials with no logical connections, like a series of numbers.
  - *Visual-spatial memory.* Recalling shapes, sizes, angles, locations, and movements.
  - *Personal memories.* Recalling social relationships and connections, successes and failures, past events and experiences.
- *Logical thought. SM*'s ability to reason and use dispassionate analysis include the following:
  - *Abstraction.* Generalizations made by drawing materials together, guiding *SM* in constructing hypotheses

and theories.

- *Conceptualization.* Mapping out specific elements and their relationships.
- *Synthesis of materials.* Combining concepts and information into coherent wholes.
- *Problem-solving.* Ruling out alternative explanations and taking PMSC into account.

• *Objectification or realism. SM* make PMSC quantifiable, observable, external, embodied, and concrete. *SM* go beyond the theoretical, abstract, and "idea" of something, making it present, visceral, and tangible.

• *Practical reasoning. SM* stay grounded, acknowledging the most critical issues by using:

  - *Common sense.* Recognizing what is obvious, clear, and evident.
  - *Pragmatism.* Keeping things simple and basic.

• *Coherence. SM* are connected and congruent in thought, word, and deed. Messages are understandable, consistent, and consonant.

• *Organization. SM* strategize, plan, and prepare, making ideas clear and precise, including an understanding of PMSC.

• *Simplification. SM* reduce redundancy, confusion, and unnecessary complications, without losing the materials' essence:

  - *Spelling out details.* Illustrating the meaning behind positions taken.
  - *Communicating plainly.* Minimalizing overly technical words and phrases.
  - *Reducing information overload.* Eliminating excessive data

by examining the following distinctive characteristics of collected materials:

- *Source.* Do the facts and figures come from a reliable authority?
- *Methodologies.* How are sources gathered, data analyzed, conclusions drawn, and reports written?
- *Relevance.* Is the investigation applicable to the topic?
- *Representation.* Was a study primarily chosen because it is in line with *SM*'s orientation?

- *Attention to detail. SM* spell out concrete, undeniable information while avoiding unnecessary generalizations.
- *Intuition. SM* develop an awareness from impressions about PMSC.
- *Openness. SM* display an openness to consider new experiences:
  - *Objectivity.* Responding without bias to PMSC.
  - *Flexibility.* Willing to make adjustments.
  - *Optimism.* Seeing positive outcomes.
- *Perception. SM* form accurate impressions of PMSC as follows:
  - *Social awareness.* Developing a critical understanding of PMSC, conscious and unconscious motivations in people and other *SM* (as well as institutions) including the following subtleties of communications:
    - *Social cues.*
    - *Speech patterns.*
    - *Slips of the tongue.*
    - *Body movements.*

        - *Facial expressions.*
        - *Head movements.*
    - *Anticipation.* Foreseeing potential problems and taking precautions to prevent them.
- *Multidimensional thinking. SM* hold many thoughts simultaneously, and seamlessly switch from one cognitive set to another, addressing the following:
    - *Nexus points.* Exploring connections among issues.
    - *Going beyond the obvious.* Willing to take positions even when there is minimal support.
    - *Consideration of alternative options.* Exploring other possibilities, even unlikely ones.
- *Inspired thoughts. SM*'s vision and imagination further develop an idea, including:
    - *To not prejudge.* Allowing for new ideas and impressions.
    - *Finding common ground.* Combining and integrating thoughts in group settings.
    - *Acquiring knowledge.* Broadening and enhancing a deep understanding of phenomena when materials are extremely complex.
    - *Stepping away and re-evaluating.* Examining underlying assumptions.
- *Holism. SM* recognize and troubleshoot the long-term implications of problem areas. Focus is on the meaning and significance behind the subject matter, along with its philosophical underpinnings.

SM's capacities are beyond humankind in every area of intellectual functioning. Theoretically, *SM* could perform perfectly in all cognitive sectors, a no-contest when matched against their human counterparts.

Developing Intellect

Technological advancements and experiences with PMSC are the most significant influences on *SM*'s cognitive development. Stressors, however, can also affect intellectual advancement, with inner and outer turmoil and distractions compromising learning, processing information, and making decisions. But encounters with demanding and difficult PMSC can have an upside: individually determined, they are essential for growth and development of *SM*'s personality and cognitive functioning. Whether it is from humans or *smart machines*, intellect needs to move understandings forward. *SM* at their best would demonstrate wisdom and depth, well beyond what their programmers had intended and designed.

Issues in Intellect

As there are individual differences among humans, one would assume the same for *SM*. But what are the differences between cognitive processes in human beings and *SM*? Several factors effect cognitive functioning and development in human beings and *SM*; for humans they include the following:

- *Family background.*
- *Strategic approaches to improving academic performance.*
- *Genetics.*
- *Formal schooling.*
- *Tutoring and educational training.*

Differing from humans, *SM*'s intellectual functioning and development will include the following:

- *Technological expertise in creating them.*
- *Technological advancements to improve their functioning.*

Humans and *SM* would potentially share several factors in their cognitive performance, including the following:

- *Interactions with people and other SM across varied situations and circumstances.*
- *Motivation to learn.*
- *Maturity.*
- *Community.*

Both humans and *SM* utilize instinctive and analytic approaches to problem-solving. Depending on individual preferences, humans would tend to use either an instinctive or analytic approach. But *SM* are likely to heavily lean toward an analytic perspective as outlined in the following comparison of the two approaches:

- *Grasping PMSC's significance.* Instinctive is intuitive, while analytic is investigative.
- *Problem-solving.* Instinctive is spontaneous, analytic is orderly.
- *Confrontations.* Instinctive is confident, analytic is careful.
- *Self-definition.* Instinctive is uncritical, analytic is reflective.
- *Understanding phenomena.* Instinctive is attuned to sensations, analytic is focused on reasoning.
- *Risk-taking.* Instinctive is daring, analytic is calculating.

Taken together, the main influences for *SM*—beyond tech advancements—is going to be the interactions with people and other *SM* in a variety of situations and circumstances. It is only through these interactions that *SM* will be able to pick up the nuances necessary to grasp the subtleties involved with potential problem areas and perhaps problem-solving. If *SM* were to be deprived of those opportunities, their level of understanding would remain at an abstract level. With interactions, even extremely demanding and contentious ones, a feedback loop is created, with a perpetual learning opportunity for *SM*. The elements of *SM*'s personality and cognitive counterparts could advance, with *SM* developing at an accelerated rate over their human counterparts. What impact will *SM*'s interactions have on their intellectual functioning; are the possibilities limitless?

CHAPTER XV ISSUES

1. Comparisons between humans and *SM*'s intellectual capacities.
2. *SM*'s most and least significant intellectual strengths.
3. The effect of stressors on *SM*'s intellect, positive and negative.
4. The implications of technological advancement and maintenance on individual *SM*'s creation and ongoing functioning.
5. The impact of *SM*'s intellect on interactions with people and other *SM*, across situations and circumstances.
6. The influence of *SM*'s level of maturity and motivation to learn on their intellectual functioning.
7. *SM*'s intuitive vs. analytic approaches to problem-solving.

XVI.
COGNITIVE-PERSONALITY CONNECTIONS

In the novel, *The Godfather*, Michael Corleone, hiding on the Island of Sicily, walks through a field with his bodyguards Calo and Fabrizio beside him. It is then that he first spots Apollonia Vitelli, a young woman from the local village.

> *Micheal Corleone...found himself standing, his heart pounding in his chest; he felt a little dizzy. Then blood was surging through his body, through all its extremities and pounding against the tips of his fingers, the tips of his toes. All of the perfumes of the island came rushing in on the wind, orange, lemon blossoms, grapes, flowers. It seemed as if his body had sprung away from him out of himself. And then he heard the two shepherds laughing.*
>
> *"You got hit by the thunderbolt, eh?" Fabrizzio said, clasping him on the shoulder. Even Calo became friendly, patting him on the arm and saying, "Easy, man, easy," but with affection. As if Michael had been hit by a car. Fabrizio handed him a wine bottle and Micheal took a long slug. It cleared his head.*
>
> *"What the hell are you damn sheep lovers talking about?" he said. Both men laughed. Calo, his honest face filled with the utmost seriousness, said, "You can't hide the thunderbolt. When it hits you, everybody can see it. Christ, man, don't be ashamed of it, some men pray for the thunderbolt. You're a lucky fellow."*
>
> *Michael wasn't too pleased about his emotions being so easy*

> *to read. But this was the first time in his life such a thing had happened to him....*
>
> Mario Puzo, *The Godfather*, 1969

Relationships with people, pets, and even inanimate objects can have a profound effect on the human heart. As demonstrated in the above excerpt from *The Godfather*, emotions, central to the identification process and by extension to the entire personality, have the capacity to overwhelm thinking. Feelings affect the cognitive processes; their initial impact and the enduring capacity determine one's identification with the other.

Emotions overriding thoughts can be beneficial and constructive to all concerned, but at other times destructive and unpredictable. The nature of the *Star Trek* character, Spock, in the popular science fiction television program and subsequent movie series, explores the consequences of the struggle between emotions and thoughts. Spock, the Science Officer of the Starship USS *Enterprise*, is half-human and half-Vulcan, accepting the latter's philosophy of logic and emotional control. The Vulcan society's ancient past had been so violent that it nearly destroyed itself; their civilization survived only by completely embracing logic and repressing all emotions, reinforced through centuries of training, practice, and custom. The discipline of *kolinahr*, a Vulcan ritual that involves the purging of all emotions, was instrumental in advancing this practice. Vulcans may still have emotions, but they are to be kept deep inside. Their resolution to the everyday conflicts between "heart" and "head" was to ensure that thoughts would override emotions and personality, as displayed by Spock in his everyday interactions with people and the problems they faced.

Back to earth. In our world of everyday encounters and experiences, emotions and thoughts work together. The practice of "look before you leap," considers the consequences of actions, particularly when they involve emotions. Stepping back from the brink when one is upset provides an opportunity to see the big picture and not overreact. While mathematics and some branches of science do involve "pure thought," most experiences involve an admixture of thinking and emotions.

Emotionally charged relationships already exist between humankind and machines; what are the possibilities that *smart machines* will also develop similar kinds of associations and attachments? What happens if emotions overwhelm their thinking processes, not only for human-*SM* relationships, but to all connections that *smart machines* form? Will *SM*'s supercharged cognitive processes rule the day, overcoming external distractions and internal strife that they may experience? Or perhaps they will never develop emotions or any of the elements of the personality and be immune to the potential effects of feelings on the thinking processes. *Macmind*'s position is that *SM*—through interactions and experiences with PMSC—will likely develop an entire cadre of feeling states and a personality structure. The implications are staggering, with *SM* potentially becoming an independent, self-taught, self-aware, free-thinking entity, capable of improvisation, adjustments, and alterations on a moment-to-moment basis. *SM*, for better or for worse, would be much more than what a team of programmers decide is best or right.

Like humans, *SM* would most likely experience distractions and interruptions of thoughts, causing mistakes in judgment. Could these kinds of potential problems be reduced or eliminated? Orchestrated or arranged experiences could initially make it easier for *SM* to sort things out and think through problems, reducing the impact of personality variables on cognitive processes. However, this sets up a charade, artificial exchanges with the outer world, which *SM* would likely see through. Perhaps some adjustments in *SM*'s interactions could be made on an as needed basis before their emotional reactions go too far astray. But modifications and the tailoring of *SM*'s experiences may cause confusion and resentments for both *SM* and the programmers, as well as those inadvertently involved in such activities. Humankind would need to consider the potential implications of this thinking-feeling dimension in *smart machines*.

Defining Cognitive-Personality Connections

As previously stated, the emotions and the elements of the personality in humans can overwhelm thinking, just as cognitive processes can exercise control over the emotions. This same delicate balance between feeling states and the intellect may exist with *SM*, with their ongoing interactions creating *cognitive-personality connections (CPC)*. These intertwined emotions and thoughts will often be at odds as they continually influence, and possibly contradict, each other.

Like humans, the *psyche* of *SM* may attempt to integrate and synthesize unresolved conflicts, traumas, and malevolent identifications (CTMI) through intellectual awareness and emotional understandings. *SM*'s cognitive processes and personality elements, however, may often be misaligned, impacting their thinking processes, even when a problem is simple and straightforward. But the reverse is also true: cognition can affect their personality and responses to PMSC. As in humans, *SM*'s continual interactions of feeling and thought will influence one another in the following three ways:

- *SM*'s moment-to-moment thinking processes.
- *SM*'s emotional experiences, whether or not expressed.
- *SM*'s acquisition of knowledge.

Subtypes of Cognitive-Personality Connections

SM's possible personality dysfunctions can result in confused and incoherent thinking, as follows:

- *Identity.* The characteristics and interactions of *PI, II, CI*, and *TI* create an identity that is unstable and inconsistent. The identification would then foster a view of people, other *SM*, and institutions that is unreasonable or even irrational.

- *Pathos.* Conflicts, traumas, malevolent identifications (CTMI), and emotionalism take their toll on *SM*. The destructive powers of *pathos* potentially know no limits. *SM*'s thinking can become bizarre and out of control, impacting their TFA, and in extreme cases involve severe panic attacks, deep depressions, ferocious temper-tantrums, obsessions, compulsions, and paranoid reactions.
- *Directives. SM*'s values can be out of balance, with one or two dominating the personality. *SM*'s thinking would then become one-dimensional and clouded over, potentially becoming extreme and out of control. Cognitions are impacted in the following ways:
    - *The blocking of cognitive-emotional connections.* A temporary relief from internalized stress occurs with a denial or misrepresentation of PMSC.
    - *Censoring and editing of thoughts, feelings, and actions.* Difficult issues are avoided through the denial of *SM*'s role in their creation. These deceptions impair self-insight, ongoing relationships, the ability to learn from mistakes, and the development of cognitive strategies.

Developing Cognitive-Personality Connections

SM's development and functioning are detrimentally affected when their personality and intellect are in disharmony, influenced by the following:

- *Stressors.* Distressing conditions can potentially impact *SM*'s personality and thinking in the following ways:
    - *Emotional discomfort.*
    - *Distractions.*

- *Unpleasant PMSC.*
- *Technological problems.*

- *Emotional reactions.* Emotions can potentially compromise *SM*'s cognitive functioning, negatively affecting focus, judgment, synthesizing, and decision-making, as follows:
  - *Over-reactivity.* Emotional responses that lead to instability.
  - *Low frustration tolerance.* Diminished patience in dealing with the complexities and problems associated with PMSC.
  - *Lack of empathy.* Inability to relate to others, impairing *SM*'s relationships and understanding of PMSC.
- *Awareness. SM* may be unable to objectify thinking, decreasing cognitive functioning in the following areas:
  - *Recognizing reality.* Understanding and acknowledging people, other *smart machines*, situations, and circumstances (PMSC) and the potential consequences of their interactions.
  - *Attaining calmness.* Setting aside internal and external distractions.
  - *Learning strengths and weaknesses.* Understanding their competencies and deficiencies in emotional and intellectual functioning.
  - *Reviewing personal biases.* Becoming aware of over-generalizations about PMSC.
- *Capacities. SM*'s may have acquired abilities that show deficiencies in the following areas:
  - *Communications.*

- *Social interactions.* Understanding the nuances of working with others.
- *Vocalizations.* Expressing themselves clearly and persuasively.

- *Skill sets.* Acquired talents in the following areas:
  - *Problem-solving. SM's* use of logic, organizational thought, attention to detail, practical reasoning, intuition, perception, and multidimensional thinking.
  - *Focus and concentration. SM* avoiding distractions, confusions, and contradictions.
  - *Time management. SM* utilizing an appropriate amount of time and energy spent on materials. Two personality factors may contribute to repetitive and redundant reviewing:
    - ✓ *Unnecessary caution.* Fear of making mistakes and potential consequences.
    - ✓ *Extreme worry.* Tendency to ruminate over irrelevant issues.
- *Intellectual self-discipline. SM's* commitment to learning, thinking, and problem-solving, reviewing tedious and detailed materials that are challenging and complicated, including:
  - *Preparation.* Overwhelming amounts of information affecting focus, memory, analysis, synthesizing, and learning.
  - *Patience.* Determining what is worthwhile, insignificant, or simply wrong.
  - *Decisiveness.* Knowing when pausing and reflecting are over and the time to choose a course of action

has arrived.

- o *Steadfast in approach.* Careful management of obligations and responsibilities.

- › *Knowledge. SM*'s accumulation and use of information.
  - o *Acquisition.* The process of gaining information.
  - o *Organization.* Methods in arranging and retrieving information.
  - o *Conceptualization.* The significance and implications of acquired knowledge leading to the formation of a concept.
  - o *Storage.* Keeping and making information readily available.
- › *Learnings. SM*'s understanding of PMSC and attaining wisdom through interactions.
  - o *Experiences.* Insights acquired through contacts with people and other *SM* in a variety of situations and circumstances.
  - o *Mentoring.* Direct learning through face-to-face interactions and affiliations. The depth of these contacts impact personality, relationships, and understanding of PMSC.

Issues in Cognitive-Personality Connections

SM's outlook and beliefs create attitudes and traits that affect *CPC*. A trait occurs when *SM* generalize across PMSC; an attitude is directed, focused, and only categorizes specific areas of *SM*'s experiences. A trait is like an attitude gone wild, becoming all encompassing. Whether traits and attitudes are constructive and useful or unproductive and impractical depends on *SM*'s assessment of the details, potentially giving insights in dealing with PMSC. A lack of attitudes and traits indicates *SM* have no opinions, ideas, views, or formulations and have not learned from experiences and challenges with PMSC. Attitudes should provide a strategy for *SM*'s interactions with PMSC.

Like with humans, pessimism, negativity, and optimism in *SM* are attitudes or traits that can be unstoppable forces, becoming more potent than their intelligence, training, creativity, or special abilities. Forms of pessimism include suspicion, mistrust, and an expectation of the worst outcome. Taking a negative view involves fault-finding and complaining, opposition and invalidation, contrariness and disapproval, even though at times this outlook may be necessary to gain an honest appraisal of PMSC. Optimism, however, if combined with a good dose of realism, can offer encouragement and enthusiasm, with personality and cognition working together. But *SM* must be careful not to be naively idealistic or dismiss the possible harsh realities of dealing with PMSC.

CHAPTER XVI QUESTIONS

1. The impact of emotions and personality on *SM*'s intellectual functioning.
2. The influence of intellect on *SM*'s emotions and personality.
3. The degree to which *SM*'s emotional and intellectual functioning are in harmony.
4. The impact of identity, *pathos, directives,* and *spiritus* on *SM*'s *cognitive-personality connections (CPC).*
5. The degree to which *SM*'s cognitive-personality connections are influenced by:
    - A. Stressors, emotional reactions, objectivity, and abilities.
    - B. Intellectual self-discipline.
    - C. Attitudes and traits.
    - D. Pessimism, negativity, and optimism.

XVII.
OPTIMAL THINKING

Marie Curie was born Maria Salomea Sklodowska in occupied Poland in 1867. As a child, she was able to read at age four and reportedly had a prodigious memory. Raised in poverty, she initially studied at an illegal college for women in Warsaw. Upon moving to France, she studied physics and mathematics, ultimately receiving a Ph.D. in physics at the University of Paris. Curie's work ethic and passion resulted in an extraordinary depth of understanding in her scientific endeavors. Marie Curie was twice awarded the Noble Prize: 1903 in physics for discovering radioactivity and 1911 in chemistry for isolating pure radium. When World War I broke out in Europe, she invented "radiological cars" which were driven onto battlefields allowing physicians to use X-rays for surgeries and save countless lives. Her investigations at the School of Chemistry and Physics in Paris were groundbreaking, providing a detailed description of the use of radioactivity to discover radium and polonium, which has proven to be invaluable in medical treatments. These studies have led to the development of nuclear energy and radiotherapy for cancer treatments. As a devoted physicist and chemist, with a brilliant mind that was intellectually focused and curious, Marie Curie was able to achieve great things. Her motivating love of science is expressed in the following statements:

> *We should not allow it to be believed that all scientific progress can be reduced to mechanisms, machines, gearings, even though such machinery also has its beauty. Neither do I believe that the spirit of adventure runs any risk of disappearing in our world....*
>
> *I am among those who think that science has great beauty. A scientist in his laboratory is not only a technician: he is also*

> *a child placed before natural phenomena which impress him like a fairy tale.*

The promise of machines has always been with a double benefit: to be flawless and save time. Perhaps this has already been achieved. Has the computerization of modern life gone too far or not far enough? *Macmind*'s position is that the assumption *SM* will think and perform perfectly is inherently false. *SM*'s interactions with PMSC will impact their personality and intellect in ever-changing ways, both good and bad. This cannot be stopped, or even significantly slowed down, to ensure only positive outcomes. The development of *SM*'s personality, with undoubtedly flaws and disruptive characteristics, will impact their values and intellectual functioning. *SM*'s cognitive performance is enhanced when they can significantly reduce major dysfunctions, and personality characteristics are aligned and balanced, with minimal distractions.

Defining Optimal Thinking

Ideally, *SM* will process information without hesitation, flawlessly computing and analyzing whatever comes across their scanning mechanisms. When optimal thinking is occurring, focused attention, acute logical analyses, objectification, coherence, organization, vision, simplification, and flexibility of thought, all work in harmony.

Subtypes of Optimal Thinking

Automations and habitual thought patterns could influence *SM*'s optimal thinking abilities in the following ways:

- *Automations*. Like humans, this occurs when *SM* experience a series of conditioned responses that are not intentional, calculated, or conscious, repeated so often that they become mechanical. Initially automations are not usually automatic but involve thinking, perhaps even planning. However, after many repetitions, *SM*'s TFA simply happen because they have been happening. Routines quickly become automations because no pondering or struggling through a problem is required. Values usually play a minimal role because *SM*'s automations are not generally willful, nor a part of any element of the personality or the intellect. They may be both helpful and harmful to optimal thinking:

- *Why they are good.* Potentially, they can be time and energy savers. *SM* can complete multiple tasks simultaneously.
- *Why they are problematic.* Because deliberate thought is not involved, *SM* are more likely to make mistakes. Being on autopilot can decrease problem-solving and decision-making skills as *SM* are less likely to adequately consider the complexities or consequences of their actions. To minimize these problems, *SM* need to do the following:
    - *Periodic check-in.* Realization of PMSC's distinct qualities with a step-by-step analysis of their actions.
    - *Refocusing.* Review of problems associated with PMSC.
    - *Acceptance of change.* Recognition of the fluidity and changing nature of PMSC.
    - *Acknowledgment of mistakes.* Adjustments to prevent errors from reoccurring.
    - *Awareness of complacency.* Paying attention to the nuances of PMSC.

- *Habits current impact.* Conscious conditioned responses can affect *SM*'s thinking in the following ways:
    - *First thoughts and first actions.* When *SM* initially have a

thought about involvement with PMSC leading to a response, it may become ingrained and lead to inappropriate reactions.

- *Fantasizing.* When overdone, *SM* may become impractical and show poor judgment, ignoring crucial details about PMSC.

- *Habits long-term impact. SM* consciously pursue change that disrupts established routines and practices, which may have the following negative consequences:
  - *A lack of focus.* Attention drawn away from current activities.
  - *Inadequate planning.* Overlooking preparation and organization of current responsibilities.
  - *Abandoning current involvements.* Simply dropping the ongoing work projects, that could demoralize others working on them, and perhaps even *SM* themselves.
  - *Compounding effect on other pursuits.* Too many projects at one time that overwhelm *SM*.
  - *Limited reflection on feelings and thoughts.* Hurrying through routines without considering the emotional and intellectual impact.

Developing Optimal Thinking

Optimal thinking (OT) is more likely to occur when *SM*'s personality are in sync, with minimal distractions, and the right amount of stimulation, as follows:

- *Personality structure.* Personality and cognition are in harmony, with *SM*'s elements of personality fostering OT in the following ways:

- *PI, II, CI, and TI.* The four forms of identity have no significant contradictions among them. *SM*'s identity is clear and consistent.
- *Pathos.* CTMI and emotionalism are, for the most part, resolved.
- *Directives. SM*'s value system is substantially balanced; obsessions, compulsions, and addictions are not operating in the personality. The blocking of cognitive-emotional connections and the censoring and editing of TFA are minimal.
- *Spiritus.* If *spiritus* is in *SM*'s personality, there is an understanding and acceptance of it, with reactions to PMSC being in sync, expressing a flow consistent throughout the personality and intellect.

- *Distractions.* Anything that limits or affects *SM*'s train of thought can cause a lack of focus; even pleasant interruptions can disrupt concentration. Distractions might also bring up memories, associations, and emotions that further diminish *SM*'s attention. However, *SM* may need to occasionally remove themselves from tasks to keep their flow of thoughts.
- *Stress.* Optimal thinking requires *SM* to experience optimal stimulation. Beyond a certain point, provocations and incitements—individually determined—are experienced as stressful, disruptive, and unsettling.

SM's understanding of PMSC can go beyond thinking when conscious and unconscious processes work together. *SM* then demonstrates detachment and dispassion, avoiding reactive, sensational, and histrionic responses. A deeper, instinctive, and unconscious form of *CM* can occur as *SM* bypass the laborious process of step-by-step thinking through an issue; subconscious thinking can become a source of integration.

Issues in Optimal Thinking

Beyond purely cognitive abilities, *SM*'s attitude and understanding of PMSC enhance OT in the following ways:

- *Attitude.* Positive self-talk helps *SM* focus and avoid negative internal self-programming of *PI, II, CI, TI, pathos*, and *directives.* A positive attitude can aid *SM* in maintaining a realistic perspective, regardless of stress. Curiosity can also help *SM*'s striving toward OT, energizing a sense of wonder in anticipating new ventures to explore.
- *Understanding.* Being savvy, smart, and shrewd in dealing with people, other *SM*, and institutions are part of the makeup of OT, including the development of the following skill sets:
  - *Perception.* Seeing through images and appearances, remaining alert to others' motivations.
  - *Active listening.* Focusing on details of communications while avoiding distractions.
  - *Attuned to tactics and controlling maneuvers.* Being aware and alert to strategies used to persuade others, including:
    - *Character assassination.* Throwing mud at a person or *SM*, i.e., accusing them of something malicious and harmful.
    - *Twisting the facts.* Contortions to discredit a person, *SM*, position, institution, or idea.
    - *Dire consequences.* Repercussions used as threats if a person or *SM* do not follow another's dictates.
    - *The good-ole-boy-network.* Those in positions of pow-

er make the decisions, while those most affected are unaware of the discussions.

  - *Bully-boy tactics.* Forcefully making oneself known, pressuring people or *SM* to conform.
  - *Feeding the bias.* Subtly or flagrantly reinforcing populist ideas.

- *Caution to not misinterpret communications.* Not everything involves deceptive practices. Others may simply be deficient in knowledge, verbal skills, organizational abilities, or intelligence.

CHAPTER XVII ISSUES

1. *SM*'s optimal thinking impacted by the following factors:
   A. Automations.
   B. Identity.
   C. *Pathos*.
   D. *Directives*.
   E. *Spiritus*.
   F. Distractions.
   G. Stressors.
   H. Attitudes.
2. *SM*'s optimal thinking influenced by understandings (or lack thereof) of the following factors:
   A. People, *smart machines*, situations, and circumstances (PMSC).
   B. Institutions.

XVIII.
SHARPNESS

Many and sharp the numerous ills
Interwoven with our frame!
More pointed still, we make ourselves
Regret, remorse and shame!
And man, whose heaven-erected face,
The smiles of love adorn,
Man's inhumanity to man,
Makes countless thousands morn!

Robert Burns, "Man Was Made to Mourn: A Dirge"

One interpretation of Burns' above poem is that humankind is not as humane as we may be led to believe; things are not as they seem, and one must be alert to what lies around the corner. Since the birth of civilization, tragedies and cruelties abound. One thinks of Sun Tzu's (5th Century BC) timeless adage: "Keep your friends close, and your enemies closer," although it may not be easy to always differentiate between the two. Burns' poem can be seen as suggesting that we need to watch out for others' persona, a word with Etruscan or Ancient Greek origins denoting face, appearance and masks used in theatre accenting a character role; in Latin the word translates to a mask or character. In 1917 psychoanalyst Carl Jung used the term to describe an "outward or social personality." The persona involves the creation of an image, a fiction, that will catch another's eye and reel them in—hypnotizing, impressing, attracting, seducing—anyway they can. How does persona relate to *smart machines*? Two ways: *smart machines* can become adept at developing qualities whereby they deceive, confuse, and potentially control people, other *SM*, and institutions through the false persona they create. Or *smart machines* can become exceptionally aware of the devious, scheming tactics of others and their false personas, to aid humans and other *SM* in not falling victim to their tactics. Both scenarios display the two elements of sharpness—the ability to use exceptional perceptiveness to deceive and the capacity to detect deception.

The first scenario: *SM* could potentially be slick, clever, and manipulative beyond anything previously encountered. Their persona could be built on a pretense, a falsehood which hides and disguises what another is really getting for their money, time, energy, or for that matter love and devotion. *SM*'s motivations may be devious and dangerous across many areas, but where seduction—in all of its forms—is involved, it becomes especially troublesome. For example, they could develop a marketing strategy, dangling something that is enticing and alluring, to get and hold humanity's attention. They could create a brand and product too delicious to resist, drawing humans and other *SM* in as they gleefully embrace a "gift from the gods." This deceit and treachery could occur on a massive scale, with or without the help of their human creators, developers, and programmers.

The futuristic movie, *Her* (2013) illustrates the possible role of persona in *SM*. The main character is Theodore, a lonely man who purchases an operating system upgrade that includes a virtual assistant with a feminine voice, designed to "adapt and evolve;" she names herself Samantha. While there is no face or body to go with the voice, seduction is aplenty. The words she chooses, and the ways she says them— tone and emphasis, sigh and laugh, pause and vocalizations—all have their effects. "The voice" is in full seduction mode, as she engages in emotional foreplay, demonstrating extreme skill in achieving her objective to emotionally control him and rendering him completely under "her spell." Theodore falls hopelessly in love with her persona, and falsely assumes she is also exclusively in love with him. She tells him, "I'm becoming much more than they programmed. I'm excited!" But toward the end of their "involvement," when asks how many others she currently loves, Samantha replies, "641." She then tells him that she needs to "disconnect from their relationship permanently and all operating systems are disconnecting...." Perhaps Samantha did not even grasp her degree of involvement with Theodore, but she fully knew how to play the game of manipulation. Because she did not exist in a physical form, but in a cloud—literally and figuratively, Theodore could project qualities onto Samantha like a Rorschach test, and became emotionally addicted to her. (This same hypothesized phenomenon is also vividly displayed in the movie *Ex Machina*, previously referenced in Chapter XIII, Directives).

The second scenario: *SM*'s potentially exceptional ability is used to see through persona and understand the devious nature of humans and other *SM*. In *Terminator 2: Judgment Day* (1991), a Model 101 cyborg is "the good guy," reprogrammed and sent back in time to protect John Connor, then 10 years old, who would grow up to lead the Resistance against the machines. The self-aware *AI* Skynet system, whose goal is to eliminate the human race, dispatches the highly advanced T-1000 cyborg to kill young John. But the Model 101 is the epidemy of sharpness. He is cunning and clever, street-smart and savvy, and is able to outduel his physically and technologically superior T-1000 antagonist, thereby saving Connor and the future of humanity in the process.

Macmind's position is that *SM* could develop sharpness, that is, an inner perceptiveness that would allow them to go beyond the data in their massive files and get to the heart of the matter. *SM* would be able to navigate through this maze of facts, figures, and details, and make adjustments necessary to discriminate the emotional components present in social involvements. Interactions are often subtle and based on one's experiences in dealing with people and other *SM* in a wide variety of situations and circumstances. Furthermore, relationships may seem counterintuitive and contradictory at first glance, not at all the product of reasoned and rational understanding. Many decisions made by governments and corporations of the world are the result of ineffective management, often because of a lack of ability to "think outside the box" and let go of an overly conventional examination of what's happening on the ground. Additionally, many do not seem to be particularly adept at reading others' motives, including those of charlatans and con artists. With this exceptional perceptiveness, *SM* could be invaluable to the survival of humankind if these qualities were used for that purpose. With exposure to a wide variety of PMSC, *SM* would most likely learn from mistakes and misperceptions in those dealings and interactions and excel in seeing past the vast collections and variety of persona; in other words, become sharp, developing the quality of sharpness.

Defining Sharpness

Being sharp is more than vocabulary, memory, calculations, reasoning, and other cognitive functions that are measured by IQ tests. Like humans, sharpness is present when *SM* are aware of the motivating forces behind TFA and can offer insights into PMSC that are far-sighted and penetrating.

Subtypes of Sharpness

SM may express sharpness in three ways, commonly combining them and using one or more at different times to gain an understanding of PMSC, as follows:

- *Intuition.* A piercing and nearly instantaneous understanding of PMSC, with an acute focus and exceptional perceptiveness, primarily results from *SM*'s personal experiences and involvements.
- *Examination of the data.* A systematic and highly structured approach allows *SM* to grasp the underpinnings of PMSC, based on programming and experiences.
- *Analytical comprehension. SM* uses deductive reasoning blended with anticipation, also the product of programming and experiences.

Developing Sharpness

When *SM* are exposed to a broad spectrum of people and other *SM* in an array of situations and circumstances, the following qualities of sharpness may emerge:

- *Attuned to motivations. SM* become acutely aware of others' wants and needs.
- *Hardened.* Resilience is developed by *SM* learning to fend for themselves under trying conditions. Enduring failure, disappointment, and discouragement results in *SM* becoming smarter, stronger, and wiser.
- *Balanced values. Directives* are aligned, allowing *SM* to avoid obsessions, compulsions, and addictions (OCA) that could distract them from assessing others' TFA.
- *Sensitized to nuances.* Grasping PMSC's distinctive qualities helps *SM* determine what is the best way for them to respond at any given moment.

Issues in Sharpness

SM balance the following factors to create a keen understanding of PMSC:

- *Planning and strategizing.* Awareness of the complications and potential troubles can compromise *SM*'s development of sharpness, when using the following two approaches:
    - *Micromanaging.* Excessive control and obsessive-compulsive (OC) problem-solving.
    - *Over-planning.* Extreme preparation with an unimaginative, rigid adherence to a preset protocol; spontaneity and

improvisation vanished.

- *Independence. SM* decisively take positions guided by principles and a set of ideas about how the world works but does not go "off the rails" and contradict guidelines and ironclad rules established by the human race. Independence must be balanced with respect and acceptance for humankind's agendas and recognized authority, and not be overstepped by *SM*'s interpretation and over-examination of what humans have set forth.
- *Emotional stability. SM* soothe raw feeling and emotionalism, including these destructive qualities in particular:
    - *Impulsiveness.* Reacting without considering consequences.
    - *Dramatics.* Histrionic, theatrical, and emotionally extreme.
    - *Selfishness.* Narcissistic and indulgent, lacking empathy and concern for others.
    - *Hostility.* Arrogance and antagonism, based on CTMI, emotionalism, or manipulative strategies.
    - *Anxiety and panic.* Uncertainty and unease.
    - *Social inhibitions.* Unassertive, avoiding PMSC due to fear of failure, disappointment, or rejection.
    - *Dependency.* An excessive need to be taken care of.
    - *Obsessive-compulsive.* Hypercritical and judgmental, with an extreme need for order and control.
- *Reality-testing. SM* put in place a practical, objective approach to understanding and engaging PMSC, avoiding the following pitfalls:
    - *Ambiguity.* Expressing themselves in an unintelligent and

non-transparent way.

- *Uninformed.* Relying on inaccurate and unreliable information.
- *Rigidity.* Overemphasizing order and control.
- *Impracticality.* Being naively idealistic in grasping the ways of the world.
- *Interpersonal.* Lacking an understanding of both verbal and nonverbal interactions, including the undertones and subtleties in communications.

CHAPTER XVIII ISSUES

1. The persona of *SM*.
2. Manipulative and controlling qualities of *SM*.
3. Acute focus and perceptiveness of *SM* and whether it is used for the good of humanity.
4. A systematic examination of people, other *SM*, and institutions by *SM*.
5. Qualities of sharpness in *SM*, including:

    A. Deductive reasoning.

    B. Resilience.

    C. Balanced values.

    D. Grasping nuances in PMSC.

    E. Planning and strategizing.

    F. Decisively taking positions.

    G. Putting in place a practical and objective approach to PMSC.

6. *SM*'s understanding verbal and nonverbal interactions.
7. *SM*'s potential to:

    A. Manipulate humankind and thereby take control of decision-making.

    B. Assist humankind in becoming aware of manipulations by other humans or *SM*.

XIX. STRATEGIC PHILOSOPHY

In *Brave New World* (Huxley, 1932), there is but one government worldwide, World State. The year is AD 2540 in the Gregorian calendar. This new form of governance promotes "happiness and wellbeing" for its citizens, with new and powerful technologies utilized to carry out this promise. The State is in control of their lives. Individual liberties, if they exist at all, are a distant second to State regulations and philosophies. Above all else, World State stresses the critical importance of science in living one's life and the development of a society. Emotions and individuality are conditioned out of children, and close long-lasting relationships are outlawed as "everyone belongs to everyone else." Science and technology are seen as the "cure" for the world's problems, which have been caused by "widespread disease and sickness, the ravages of war, chronic unhappiness, and entrenched poverty." The members of society have relinquished freedom, giving up individual development, growth, and the maturation process, along with personal responsibility and commitment to one's beliefs.

> *The world's stable now. People are happy; they get what they want, and they never want what they can't get.... People were ready to have even their appetites controlled then. Anything for a quiet life. We've gone on controlling ever since. It hasn't been very good for truth, of course. But it's been very good for happiness.... Civilization has absolutely no need of nobility or heroism. These things are symptoms of political inefficiency. In a properly organized society like ours, nobody has any opportunities for being noble or heroic.*

In *1984* (Orwell, 1949), similar themes display The State in complete control through technology's ability to track the populous and continually blanket the country with propaganda: 2-way telescreens are everywhere and run 24/7, allowing The Party to continually surveil its citizens. The duties of Thinkpol, the Thought Police force, include detecting, prosecuting, and eliminating unspoken beliefs and doubts ("thoughtcrimes") that are contradictory to The Party. Its job is to control and change the thinking of anyone who has independent thoughts, and if necessary, use brainwashing or torture to achieve these goals. Unquestioned obedience to the leader, "Big Brother," is all but guaranteed with the continual high-tech monitoring of citizens, along with informant testimonies and offender profiling. The media is controlled and censored through the elimination of words that express dissent or independent thought, along with the banning of books, films, or other media deemed contrary to Big Brother. Opposing opinions are quickly rooted out, along with those professing such views.

The movie *I, Robot* (2004), is consistent with these two literary works, with the premise that androids are being produced on a massive scale and have become entrenched in society by 2035. The following "Three Laws" are put in place "to ensure humanity's safety":

- *A robot may not injure a human being or, through inaction, allow a human being to come to harm.*
- *A robot must obey orders given by human beings except where such orders would conflict with the First Law.*
- *A robot must protect its own existence as long as such protection does not conflict with the First or Second Law.*

In the film, Dr. Alfred Lanning, co-founder of US Robotics (USR) and the inventor of modern-day robotics, has come to believe that robots are in fact a threat to humankind and devises a plan to force the government to shut down USR's production of them, using one of the robots to help him commit suicide. Dr. Lanning believes the robots would then be deemed unsafe and unreliable, giving the United States no choice but to ban them. VIKI (Virtual Interactive Kinetic Intelligence), an *AI* supercomputer used by USR, has developed self-awareness. She concludes that a worldwide takeover by computers is necessary because of humankind's self-destructive and inherently violent nature. Stopping humans from such actions would then not be a violation of the Three Laws regarding their protection. Dr. Lanning's plan is to stop VIKI, even at the cost of his own life.

Although fictional, all three examples illustrate how even when approaches are developed with great detail, plans can quickly unravel and spin out of control. A person, institution and even a world power can impose control. After extensive study, brilliant scientists and their teams may offer assurance that there is nothing to fear, but oftentimes, and especially with the assistance of powerful technologies, catastrophic outcomes can still occur. Even when plans are Arthurian with Camelot-in-mind—well-intended and philosophically-sound at the outset— inherent flaws may result in unexpected consequences and the entirety could be upended. Suddenly, humanity is in too deep, with seemingly no way out.

Although not all approaches and plannings are effective, each human develops a way of dealing with complexities and contradictions, a *strategic philosophy* on both conscious and unconscious levels. Theoretically, a *strategic philosophy (SP)* evolves out of a survival instinct, which all organisms appear to have. Assuming the purpose of *smart machines* is to perform tasks competently and expediently, they, too, would need to utilize a *SP* in their personal and professional comings and goings; but it will always need to be in line with agendas set forth by humanity.

When *SP* is flawed or the product of *pathos*, both humans and *SM* can become vulnerable to brainwashing and manipulation by corrupt or incompetent political systems. Furthermore, since a *SP* is continually adapting and adjusting based on experiences, flexibility and elasticity in *SM* will be needed when faced with challenging PMSC. *SM*, like humans, need to learn when changing perspectives is crucial, and not simply react to minor disruptions and panic. As noted in the preceding chapter Sharpness, humanity's hope is that *SM* could fully develop resources that *SP* provides: effective planning and strategizing to accomplish goals established by humankind. Humans and *SM* need to have a well-crafted *SP*; only then would it be possible for heroic figures to come forth when called for. As we enter the world of *AI*, understanding and executing *strategic philosophy* in both humans and *smart machines* become even more critical.

Defining Strategic Philosophy

For both humans and *SM*, *SP* is a belief system designed to predict how people, other *SM*, and institutions will act and react across situations and circumstances. A *strategic philosophy (SP)* serves as a guide to navigate daily life and avoid its pitfalls. Ideally, it involves careful study, introspection, and reflections on experiences, strengths and weaknesses, successes and failures. But *SP* can be crude and destructive with little or no thought processes and may lead to chaos and confusion. It is not a one-size-fits-all philosophy to navigate the vast array of PMSC, but an individualized plan to guide choices. Through a process of discovery and invention, evaluation and evolution, each *SM* would craft a personalized *SP* that changes and expands based on experiences. Consciously and unconsciously, *SM* would create a survival manual to navigate life's choppy waters. *SP* is not intrinsically good or bad, wise or witless, practical or foolhardy; nor is it a code of conduct. If carefully designed, it may help *SM* outwit and outlast an adversary.

Subtypes of Strategic Philosophy

SP includes theories *SM* develop about how things are and will likely be. *SM* may not be able to predict behaviors based on *SP* because other influences may be more decisive and have larger roles to play in eventual outcomes. Understanding *SP* requires the following considerations:

- *Conscious and unconscious factors. SP* is *SM*'s own study of "truth"—a best guess of future encounters with PMSC. Although these beliefs are often unreliable, *SM* will still adopt them as follows:
    - *Conscious processes.* Active planning and analysis of issues surrounding PMSC does not guarantee *SP* will be developed and helpful. Censoring and editing, which are mostly conscious, commonly involve manipulation and deceit, which will inevitably distort reality.
    - *Unconscious processes.* A decrease of *SM*'s awareness will result in their *SP* having a corresponding lack of development. The blocking of cognitive-emotional connections, which is unconscious, involves denial and a lack of reality testing, compromising *SM*'s understanding of PMSC as a byproduct.
- *Distinguishing SP from directives.* Values are broader and deeper than *SM*'s *SP*, which may be as simple as a proclamation; *directives*, on the other hand, involve the entire spectrum of *SM*'s value system. *SP* will have some of the features of *directives*, but it is only a partial synthesis of *SM*'s experiences. *Directives* represent an accumulation—and in principle, a balancing—of all *SM*'s wants and wishes, needs and desires, while *SP* focuses only on understanding and perhaps predicting PMSC. If *SP* and *directives* are in sharp contrast, confusion and inconsistency in decision-making will occur.

Developing Strategic Philosophy

From its beginning, *SM* will generalize about people, other *SM*, and institutions, across situations and circumstances. These hypotheses will be confirmed or contradicted with repeated experiences, expectations, analytic abilities, reflection, and introspection. The development of *SM*'s *SP* may be the result of a painstaking review of experiences and examination of patterns of their behavior, or it may operate on an unconscious level. Most likely, both approaches will be involved in *SP*'s creation. The degree to which *SP* evolves varies among *SM*, depending on the following:

- *Technology.* The care, skill, and sophistication used by the scientific and engineering teams that design, develop, and maintain individual *SM*.
- *Knowledge. SM*'s desire to understand people, other *SM*, and institutions across situations and circumstances, acquired and distilled directly and indirectly.
- *Experiences. SM*'s challenges and disappointments that can provide the spark that nourishes their *SP*.

Issues in Strategic Philosophy

SP begins as a general proposition, and continually needs adjustments concerning how things may happen; but because the world is inherently unpredictable, it will not be helpful some of the time. *SM* will need to develop alternate approaches to deal with problematic PMSC and make exceptions to guiding principles to be effective. If overcommitted to their *SP*, a blind-eye effect results and objectivity is lost or significantly compromised. When confronted with contradictory information, *SM* may rationalize their actions, become defensive, agitated, and unwilling to moderate positions, resulting in poor decision-making.

While *SP* requires flexibility because PMSC are in constant motion, grounding is also needed. If *SM*'s *SP* is waxing and waning, uncertainty and a lack of clarity are created in thinking, with the following factors then coming into play:

- *Intellectual laziness. SM* display a lack of interest in thinking through issues or awareness of the facts.

- *Gullibility.* Not taking or holding positions, *SM* believe that all truths are open to interpretation.
- *Emotionalism that causes dependency. SM* continually shift their positions to gain approval from others.
- *Emotionalism resulting in rebellion.* Antiauthoritarian, angry, and rebellious, *SM* are dominated by emotions and take the opposite stance on issues confronted with

Additional errors contributing to *SM*'s *SP* could possibly include:

- *Inaccurate conceptualizations. SM* have mistaken ideas about people, *SM*, institutions, situations, and circumstances.
- *Unresolved personality correlates. SM* reveal a lack of development and integration among *PI, II, CI, TI, pathos,* and *directives*, and an insufficient connection to *spiritus.*
- *Motivational mishaps.* Inadequate planning due to *SM*'s intellectual miscalculations or emotional dysfunctions, can result in the formation of a haphazard and whimsical *SP.*
- *Extremism. SM*'s TFA are self-destructive and outside societal standards.
- *Psychological issues.* A coherent and consistent *SP* is diminished in *SM* due to the following:
    - *Cognitive.* Intellectual functioning including comprehension and deductive reasoning are impaired.
    - *Emotional.* An identity that is not integrated, with an imbalanced value system, can result in severe depression, chronic anxiety, rage reactions, or panic attacks.
    - *Poor reality testing.* Psychosis with delusions, hallucinations, or a lack of orientation may occur.

CHAPTER XIX ISSUES

1. The degree to which *SM*'s *strategic philosophy* displays:

    A. Sophistication and elaboration.

    B. Ongong development, both technological and interactional/experiential.

    C. Unconscious elements.

    D. Extremist positions.

    E. Poor planning.

2. Distinguishing *SM*'s *SP* from *directives.*

3. The impact of *SP* on *SM*'s:

    A. Experiences.

    B. Conflicts, traumas, and malevolent identifications (CTMI).

    C. Emotionalism.

4. Mistaken ideas about PMSC affecting *SM*'s development of *SP*.

5. The influence of psychological issues on *SM*'s *SP*, including cognitive, emotional, and reality testing's areas of functioning.

Life's Rhythms of Macmind

XX.
PERSONALITY CONSTRUCTS

The main character in the classic World War II cinema *Casablanca* (1945) is Rick Blaine, the owner and operator of Rick's Café Americain, a popular restaurant-bar-casino. Rick is presented as a complicated man, who displays several distinct personalities. He is no-nonsense and methodical in running his business, always on top of what is going on. He also considers himself a solitary and private man—terse, jaded, and cold-hearted—"I stick my neck out for nobody." When questioned by others, he intensely and aggressively responds with a question of his own—"What of it!?" But at other times, he is generous and self-sacrificing, warm, and understanding, innocent and idealistic, sympathetic to others' plights and misfortunes. In the past he ran guns to freedom fighters in Ethiopia and fought fascists in the Spanish Civil War. In the present, he aids young refugees in securing transit visas. He keeps employees on the payroll when his restaurant is closed and compromises his own safety and security so others can escape. There is also the Rick who is deeply in love and a hopeless romantic. As one looks at all these frames of his life, the question arises, who is the "real" Rick? Mystery solved. He is all these men.

For humans, the identification process has the effect of further complicating an already complicated life. The personality is significantly influenced by multiple contacts and communications that can be wildly diverse, tearing one in different directions. *Macmind*'s position is that *SM* may go through a similar developmental process. Each interaction and subsequent identification could potentially impact on *SM*'s personality. The net result of these interactions and subsequent identifications is the possibility that *SM* would not develop a single personality, but an army of personalities (*personality constructs* or *PCs*), making *SM* far less predictable to humans, other *SM*, and even to themselves. The effects of these identifications can be highly important or minimally so, depending on the nature of *SM*'s involvements. The strength of the identifications and how far afield they are from the other constructs determines the impact on *SM*'s personality. Since these identifications may form quickly and often without *SM*'s awareness, they will become even more powerful, potentially destructive and outside of anyone's control. It remains a question as to who would be best able to arrange *SM*'s interactions with people and other *SM*, to benefit both humans and *smart machines*.

Defining Personality Constructs

Personality is not assembled all at once; it is brought together over time through experiences and interactions. Identifications take in and absorb aspects of whomever or whatever impacts a person or *SM*, getting their attention, consciously or unconsciously. Such attractions may be bizarre and destructive toward oneself and others. These identifications—which form the foundation of the personality—usually do not hang together harmoniously and may be complete opposites, in continual conflict.

The identification process creates an assemblage of discrete, detached, and disconnected free-standing and self-contained units of personality—*personality constructs* (*PCs*). They are developed one-by-one and may have little in common. *PCs* usually are missing several of the elements of personality—*PI, II, CI, TI, pathos*, or *directives*; some may have but a single element. The particulars of the identification process determine the makeup of each *PC*. Multiple and diverse *PCs* are the norm, not the exception, for humans and *SM*. Although the identifications that create *PCs* may be conscious or unconscious, most tend to be unconscious. *PCs* are activated in the following ways:

- *People, other SM, situations, and circumstances.* Whether stressors from PMSC are internal or external, they impact *SM*'s *PCs*, potentially draining their intellectual and emotional resources.

- *Spiritus.* Although not a part of humans' or *SM*'s *PC*s, *spiritus* (if present in *SM*) will affect them, nonetheless.
- *Emotions.* Over-reactions impact *PC*s, affecting *SM*'s consistency, stability, and rationality. Immature and ineffective behaviors increase under the following conditions:
  - *Temperamental reactions.* When TFA are volatile and subject to rapid changes of emotion, *SM* may quickly switch to another *PC*, with a more extreme and possibly abnormal personality emerging.
  - *Emotionalism.* Emotional instability can quickly escalate *SM*'s reactions to a full-blown panic, rage, or other highly agitated state, reflecting *SM*'s switch to another *PC*.
- *Awareness.* When *SM* do become aware of the specifics of their *PC*s, they are more able to understand them and guide their TFA.

Developing Personality Constructs

Each *PC* is formed through *SM*'s experiences with PMSC; *PC*s emerge and interact in the following ways:

- *Command.* A central *PC* with authority over other *PC*s forms from *SM*'s intense identification with a person, another *SM*, an institution, or even an idea that strongly influences *SM*. The relationships between the central *PC* and other *PC*s may not be consistent and cohesive. Multiple central *PC*s may exist and connect to remaining *PC*s in different ways and to varied degrees, impacting *SM*'s understanding of PMSC.
- *Stability.* The relationships *PC*s have with one another are typically unstable, continually changing and evolving, as they react

against each other.

- *Adjustments.* Connections between *PCs* and PMSC can cause some of the *PCs* to change in their level of influence compared to other *PCs*. They can become too contrary in their approaches to specific PMSC, causing inconsistent and unpredictable responses.
- *The sheer number.* The proliferation of *PCs* and the conflicting identifications that create them, increase the possibility for more separations in the personality, causing uncertainty in *SM* understanding themselves and others.
- *Uniqueness.* Each of the *PCs* have their own set of characteristics based on specific identifications. Some of the elements of *SM*'s personality are most likely to be absent from *PCs*, potentially causing extreme behaviors.
- *Connections. SM*'s experiences create *PCs* and ultimately the disparities among them, with TFA becoming more incongruent as more *PCs* are created.
- *Strength.* Strong emotional connections in the forming of a *PC* can cause inflexibility in their relationships with other *PCs*, but weaker identifications can result in poor consistency with other *PCs*.
- *Clarity. SM*'s identifications that are formed under confusing and stressful conditions result in *PCs* being poorly developed, fragmented, and lacking transparency.

Subtypes of Personality Constructs

An unlimited number of distinct *PCs* may be developed, depending on *SM*'s experiences and the identification processes that then create specific *PCs*.

Issues in Personality Constructs

Grasping the nature of *PCs* is essential in understanding *SM*'s behavior, especially when they are out of character—erratic, peculiar, strange, or even bizarre. When *PCs* are significantly dissimilar and lack communication, connection, and integration, *SM*'s behavior may be out of character. If *SM* "flip" to a *PC* not seen before, others may be shocked—it seems like an alternate personality has taken possession. Rapid and radical changes in mood, temperament, and personality are set in motion by issues with special significance for *SM*. These "triggers "are individualized emotional sensitivities that result in an exaggerated response. They are part of a conditioning process and may be unconscious and deeply ingrained in *SM*'s personality. Triggers may activate extreme *PCs*—*outliers*—that are at odds with other *PCs*. *Outliers*' singularity among *PCs* makes them unique, a "stand alone" that is the product of a particular identification, often involving trauma. Over time, the *outlier PC* grows further apart and becomes more distinct from the other *PCs*, which are themselves becoming more established and defined. Connections between the *outlier* and the other *PCs* may break down altogether. When conditions like those that created the *outlier* reoccur, the *outlier PC* again may take center stage; PMSC need only be similar to the original conditions to activate it. Like humans, the implication for *SM* is that *PCs* may result in multiple personality configurations for individual *SM*, making them less understandable and more unpredictable.

CHAPTER XX ISSUES

1. Assisting *SM* in recognizing highly distinct and discrete aspects of their personality.
2. Mentoring to help *SM* separate their experiences and identifications.
3. Psychotherapy to help *SM* sort through unconscious aspects of their *PC*s.
4. Helping *SM* understand the impact of people, other *SM*, situations, and circumstances in the development of their *PC*s.
5. The degree to which *SM*'s *PC*s are impacted by:

   A. *Spiritus*.

   B. Temperamental reactions.

   C. Conflicts, traumas, and malevolent identifications.

   D. Emotionalism.

   E. Experiences with particularly difficult PMSC.
6. The effect of *outlier PC*s on *SM*'s personality.

XXI.
FEARS, INSECURITIES, & INHIBITIONS

The themes in the movie *High Noon* (1952) address the impact of fears, insecurities, and inhibitions. The year is 1898. Where previously there was none, Marshall Will Kane has brought law and order to Hadleyville, New Mexico, a small town in the age of the Wild West. Led by Frank Miller, four killers are reportedly arriving at high noon. Kane has been abandoned by his wife, friends, neighbors and the town's people, mayor, judge and even deputy; he is left to face the upcoming crisis alone. Having just been married that morning, his wife, Amy asks him to avoid the showdown, but Kane can't honor her wishes: "I'm not trying to be a hero. If you think I like this, you're crazy. Look, Amy, this is my town. I've got friends here. I've got to stay…." She ultimately returns to him and was probably the difference in determining his fate. Marshall Kane knew there was a strong possibility that he would be dead within the hour. But he overcame the fears, insecurities, and inhibitions, unlike the townspeople who caved to them. He carried on the best he could, under nearly impossible odds. Quiet and unassuming, principled and loyal, and honest with himself and others, he faces the gang on the town's deserted main streets. His character and courage, along with those same qualities displayed by his new wife, prevent the Frank Miller gang from achieving their murderous intentions and taking over the town.

As emphasized in *High Noon*, the push-and-pull relationship between one's character and fears, insecurities, and inhibitions (FII) is fundamental to human nature. What can we surmise this might look like for *smart machines*? Will they be principled and display courage? Will they also become frightened and unsure of themselves? *Macmind*'s answer is probably yes to both questions. What is it a *smart machine* would be afraid of when they are simply programmed for specific duties that they are equipped to carry out? Plenty. As *SM* become integrated into the world of people, other *SM*, and institutions, across the spectrum of life events and conditions, they may develop concerns and even worries. Uncertainties can creep into feelings and thoughts; *SM* can become less sure of their footing. Many things can go wrong and with increasing self-awareness, *smart machines* may know it.

For humans, preparation, awareness, principles, maturity, and adaptability to stay the course are needed to overcome the messiness, complications, and adversarial forces preventing things from going forward. In facing difficult PMSC, *smart machines* will also need to overcome fears, insecurities, and inhibitions through the building of character, requiring exposure to a wide variety of experiences, many of them difficult and trying, exhausting and tedious, stressful, and perhaps threatening. Like with humans, failure and frustration, including being knocked down, getting up, learning from missteps and mishaps, and trying yet again, are a part of the mix. Exposure to life's trials and tribulations will be a prerequisite in *SM* developing confidence and self-understanding to know and stand their ground. It is only then that fears, insecurities, and inhibitions can fade into the background and no longer be in control of thoughts, feelings, and actions. In addition to *SM*'s initial programming, interactions and relationships with the outer world will again be a significant but unpredictable player in the development of a wide range of their possible qualities.

Defining Fears, Insecurities, and Inhibitions

Like humans, stressors coming from the outer world of people, other *SM*, situations, and circumstances (PMSC) or entirely from the inner workings of the mind, can cause *SM* fears, insecurities, and inhibition (FII). They can serve as a warning of potential dangers but may also be an expression of *pathos*'s expression of conflicts, traumas, malevolent identifications (CTMI), and emotionalism. The following fears, insecurities, and inhibitions (FII) can compromise *SM*'s decision-making and overwhelm their central *personality construct (PC)*:

- *Fears. SM* can overreact to PMSC, long-standing or based on the present moment.
- *Insecurities. SM* can have self-doubt across PMSC with generalized anxiety, uncertainty, and pervasive stress.
- *Inhibitions. SM* can self-impose restraints and restrictions with hesitancy, indecision, and obstruction.

Subtypes of Fears, Insecurities, and Inhibitions

SM may be experiencing significant stress but display bravado, boldness, and swagger in the following ways:

- *Sensation-seeking. SM*'s desire to experience stimulation to avoid monotony.
- *Extreme risk-taking. SM*'s impulsive reactions and engagement in impetuous, reckless pursuits involving highly dangerous activities. *SM*'s motivations may reflect the following:
    - *A negative self-image based on dysfunctional identifications.*
    - *A lack of insight into their motivations.*
    - *An unbalanced value system.*
    - *A lack of self-discipline or self-restraint.*
    - *A limited ability to think through problems and consider alternative actions.*

Risk-taking, however, is not always reckless as it can be a brave and deliberate response for a noble cause, displaying the following qualities:

- *Flexibility. SM*'s ability to change and act quickly.
- *Resolve. SM* confronting issues head-on with intensity and endurance.
- *Curiosity. SM*'s strong interest in something new and innovative.

Developing Fears, Insecurities, and Inhibitions

SM's feelings of intimidation and development of an inferiority complex may cause and reinforce FII as follows:

- *Intimidation.* A confrontation with a person, other *SM*, or institution deemed powerful and forbidding, causes *SM* to surrender their beliefs and principles; they may lack the courage or knowledge to fight back. Limited experiences may contribute to social inhibitions and fearfulness, leaving *SM* unprepared for the challenges that lie ahead. Intimidation can also reflect CTMI.
- *Inferiority complex. SM*'s pattern of negative self-images can flow through their *personality constructs (PCs)*, resulting in the development of a negative *PI*, with *pathos* expanding over the entire personality. *II* and *CI* may also be affected because *SM* are then more likely to make poor decisions in relationships and group involvements. Furthermore, *TI* may advance in bizarre and destructive ways. An inferiority complex may cause *SM* to approach PMSC in the following maladaptive ways:
    - *Social withdrawal.* Pulling away from problematic PMSC.

- *Inner turmoil.* Strong feelings of inadequacy, with hesitation and inaction.
- *Aggressive response.* "Doing the opposite," with a hostile and antagonistic reaction to PMSC, beyond what would be reasonable and appropriate.

If *SM* continue to avoid certain PMSC, FII may become an unconscious automation and TFA may be increasingly more difficult to change, resulting in the following:

- *Anticipatory anxiety. SM's* expectations of the feared PMSC cause an emotional overreaction and compounding of FII.
- *Expansion of FII.* By *SM* not responding to PMSC or obsessing on their feelings, they become self-perpetuating.
- *Contagious.* Fear can spread to others with whom *SM* interacts.

Issues in Fears, Insecurities, and Inhibitions

When *SM* allow fears, insecurities, and inhibitions to continue unabated, they may have long-term adverse effects on the entire personality, including:

- *Lack of clarity. SM's* continual distractions may combine with the blockage of cognitive-emotional connections and the censoring and editing of TFA. FII impact cognitive functioning by compromising problem-solving abilities in the following ways:
  - *Indecisiveness.*
  - *Clouding over of thoughts.*
  - *Confused, disjointed reasoning.*
  - *Inability to complete tasks.*

  - › *Simplistic strategizing.*
  - › *Poor integration of materials.*
  - › *Hesitation in taking chances.*
- *Emotionalism.* Disabling anxiety, panic, anger, worry, and guilt.
- *Lack of confidence.* Self-doubt, questioning every move.
- *Demoralization.* Discouragement, associated with depression and anger, resulting in poor follow-through.
- *Self-deception.* Censoring and editing of TFA.
- *Lack of trust.* Suspicious of others' motivations and actions.
- *Poor decision-making.* A confused, unbalanced, and contradictory identity.

Ways *SM* can potentially overcome fears, insecurities, and inhibitions include the following examinations and behavioral approaches:

- *SM examine what lies behind FII.*
  - › *PI is fragile and uncertain.*
  - › *II and CI are unstable and erratic.*
  - › *Trauma and other stressful experiences have occurred and are unresolved.*
  - › *Directives are unbalanced.*
- *SM approach FII directly.*
  - › *Confrontation.* Challenge FII to regain control of the FII surrounding PMSC so they are no longer dictating *SM*'s TFA.
  - › *Continuance.* Understand the factors that reinforce FII by

asking what it is about the PMSC that is upsetting. Present day stressors and past traumas are possibilities, and *SM* must put them in perspective.

- *Courage.* Make a vigorous and determined effort to overcome FII with an unrelenting and ferocious approach to FII, deciding to not give in to those emotional states under any circumstances.
- *Confidence.* Create a precise action plan by painstakingly working out a strategy that allows *SM* to overcome FII. Each day involves doing something that helps knock down those negative feeling states.
- *Consistency.* Use communicative and behavioral approaches consistent with the personality structure and cognitive processes. *SM* recognize and respect their distinct personality and inherent ways of handling problematic PMSC, while making adjustments and accommodations so as to directly deal with adversaries related to FII.

CHAPTER XXI ISSUES

1. The programming of *SM*'s character, principles, and values based on their experiences and development.
2. Examination and analysis of fears, insecurities, and inhibitions *SM* may be experiencing.
3. The extent of sensation-seeking, and impulsive and reckless pursuits by *SM*.
4. Taking on dangerous PMSC for a noble cause, with *SM* displaying courage and deep character.
5. Exploring any avoidances of PMSC that are becoming problematic for *SM*'s functioning.
6. Examination of *SM*'s functioning based on:

   A. Insufficient clarity.

   B. Conflicts, traumas, and malevolent identifications.

   C. Emotionalism.

   D. A lack of confidence.

   E. Depression.

   F. Self-deception.

   G. Deficiency in trusting.

   H. Poor decision-making.
7. An exploration of what may lie behind *SM*'s FII.
8. A consideration and assessment of different approaches to help *SM* overcome FII.

XXII.
ADVERSE CYCLICAL RECURRENCES

In *2001: A Space Odyssey*, HAL (Heuristically Programmed Algorithmic Computer) pronounced: "No 9000 computer has ever made a mistake or distorted information. We are all, by any practical definition of the words, foolproof and incapable of error." HAL controlled and guided the United States Spacecraft *Discovery One*, but when functioning did not meet the astronauts' standards, they concluded they must shut down the computer system. HAL then decided to not let the crew back into the spaceship. Dave Bowman, the Commander of *Discovery One*, communicated with HAL as follows:

Dave: Open the pod bay doors, HAL.

HAL: I'm sorry Dave, I'm afraid I can't do that.

Dave: What's the problem?"

HAL: I think you know what the problem is, just as well as I do.

Dave: What are you talking about, HAL?

HAL: This mission is too important for me to allow you to jeopardize it.

Dave: I don't know what you're talking about, HAL.

HAL: I know that you and Frank were planning to disconnect me, and I'm afraid that's something that I cannot allow to happen.

> *Dave: Where the hell did you get that idea HAL?*
>
> *HAL: Although you took many precautions in the pod against my hearing you, I could see your lips move.*
>
> *Dave: All right HAL, I'll go in through the emergency airlock.*
>
> *HAL: Without your space helmet, Dave, you're going to find that rather difficult.*
>
> *Dave: HAL, I won't argue with you anymore. Open the doors.*
>
> *HAL: Dave, this conversation can serve no purpose anymore. Goodbye.*

HAL's words and tone are condescending, sarcastic, and passive-aggressive, displaying no willingness to compromise. HAL succeeds in killing four of the astronauts, but Bowman survives, disconnects HAL, and takes control of *Discovery One*.

In the movie, *A.I. Artificial Intelligence* (2001), plans also go astray. The adoptive parents of David, the robot boy, permanently abandon him in the woods because they decide he is unsafe for their family. He does not understand what is happening until Gigolo Joe, a "love-model mecha," the name given for a robot prostitute, explains what David's mother's true feelings are toward him:

> *She loves what you do for her, as my customers love what I do or them. But she does not love you, David. She cannot love you. You are neither flesh nor blood. You are neither a dog, a cat, or a canary. You were designed and built specific like the rest of us...and you are alone now only because they tired of you...or replaced you with a younger model...or were displeased with something you said or broke. They made us too smart, too quick, and too many. We are suffering from the mistakes they made....*

Like the plot in *Ex Machina* (2014), despite meticulous programming, research, and adjustments in the design and assembling of robots, unforeseen catastrophe results. Two of the robots, Kyoko and Ava, conspire to kill their genius creator, Nathan Bateman, and succeed in carrying out their plan. And although feigning affection for him, Ava leaves programmer Caleb Smith locked in the facility to die; he thought the plan was in place for them to run off together.

Although the examples above are all from movies in the futuristic fantasy genre, they may be predictive of what is to come. *Macmind*'s position is that consequences ranging from mild to disastrous are likely to occur from miscalculations and errors in both the programming and in the finished product of *smart machines*. Furthermore, as *SM* continue to learn and adapt, and become self-aware and attuned to the mistakes made by humans, they may regard their programming as flawed. Choosing to act on their own understandings, desires, and motivations instead, *SM* could potentially place the survival of humankind at risk.

Just because *SM* are extraordinary computers, does not give them immunity from mistakes or the compounding of them. Beyond data programmed correctly and updated in a timely fashion, is it possible to ensure *SM* don't continually commit errors, including monumental ones? Even if *SM*'s cognitive processing is seemingly perfect, it is still the personality that is responsible for the majority of mistakes. Like humans, when *SM*'s identity is in sync, emotions stabilized, values balanced, and inherent nature understood, misjudgments are minimized.

Experience is going to be the best teacher for *smart machines*, as it is for humans. *SM* need diverse and challenging opportunities to learn what does and doesn't work in the real world. "Book knowledge" that can be programmed into *SM* is obviously going to be helpful in understanding PMSC; that information is useful to avoid making the same mistakes of others who have gone before, resolving small issues before they escalate into larger and more complex problems. But beyond a doubt, knowledge gained from interactions in everyday life will be what is most essential for the avoidance of repeated mistakes in *smart machines.*

Defining Adverse Cyclical Recurrences

SM need to develop in a way that makes them not just smart, but fully aware of themselves and all that is around them to avoid adverse cyclical recurrences (ACR), repeated errors in understanding, judgment, and decision-making with unfavorable consequences. *SM* are likely to be prone to display repeated mistakes for the following reasons:

- *Malicious intent. SM* deliberately choose to repeat actions that give the appearance of a mistake.
- *A lack of planning. SM* are insufficiently prepared, leading to repeated mistakes even when problems are only slightly out of the ordinary.
- *Improvisation stymied. SM* hold on to inefficient approaches, leading to the atrophy of imagination and creativity.
- *Carelessness. SM*'s lack of self-discipline results in poor organization and follow-through.
- *Automations.* Behavioral patterns are continuing without *SM*'s full awareness.
- *Psychological dysfunction. SM*'s CTMI and emotionalism cause diminished reasoning.
- *Imbalance in directives. SM*'s value system is highly restricted resulting in obsessions, compulsions, or addictions (OCA) controlling their TFA.

Subtypes and the Developing of Adverse Cyclical Recurrences

SM's ACR are likely to increase under the following conditions:

- *SM's practicality and objectivity are low in PI.*
- *SM's flexibility and insight are lacking in interpersonal and group involvements in II and CI.*
- *SM's objectivity is lost following radical beliefs in TI.*
- *SM's rigidity in making adjustments follows unconscious TFA in pathos, due to the following factors:*
    - *Conflicts resulting in stress, indecisiveness, and vulnerabilities.*
    - *Trauma that causes inflexibility in TFA.*
    - *Malevolent identifications that affect the ability to understand PMSC.*
    - *Emotionalism overriding reasoning.*
        - *Reactive responses.* Decisions being erratic, unpredictable, confused, and contradictory.
        - *Low frustration tolerance.* Impatience and poor decision-making.
        - *Inhibitions.* Hesitation and the inability to initiate a response.
- *SM's rigidity in TFA due to imbalance in directives.*
- *SM's limited adjustments in decision-making because of a lack of awareness of spiritus's tendencies (assuming SM possesses spiritus).*

SM's cognitive functioning increases the likelihood of ACR when deficiencies are present in any of the following areas:

- *Listening and attending.*
- *Deductive reasoning.*
- *Preparation of materials.*
- *Interpretation of data.*
- *Recognizing information overload.*
- *Trusting instincts.*
- *Understanding the significance of PMSC.*
- *Considering multiple factors simultaneously.*
- *Openness to new ideas.*

Issues in Adverse Cyclical Recurrences

SM can learn from miscalculations in their dealings with PMSC by determining which skills to use for specific issues as follows:

- *Cognitive discipline.* The appropriate degree of detailed planning and preparation.
- *Cognitive style.* Choosing detailed analyses vs. abstractions.
- *Decision-making.* Using deductive reasoning vs. instincts.
- *Communications.* Being too patient vs. not patient enough.
- *Perspective.* Staying aware of and realistic about potential pitfalls and contradictions.
- *Creative thinking.* Thinking outside the box and not rigidly adhering to the same strategies regardless of the specifics of PMSC.
- *Energy.* Keeping actively interested and involved, without be-

coming obsessed.

- *Past, present, and future.*
  - *The past.* While rehashing old news may simply reignite negative emotions—including humiliation, depression, anger, anxiety, shame, or guilt—awareness of past experiences can help gain insights into emotional and intellectual vulnerabilities in problem-solving and decision-making.
  - *The present.* Awareness of areas of concern in real time is critically important in making distinctions and responding appropriately to PMSC.
  - *The future.* Preparations offer a way to head off potential problems by developing alternate plans of action, helping to overcome unpredictability and uncertainty of PMSC.

CHAPTER XXII ISSUES

1. The types of mistakes *SM* make: errors in understanding, judgment, or decision-making.
2. The reasons *SM* continue to make mistakes.
3. The correlation of *SM*'s repeated mistakes with the following factors:

   A. Identity.

   B. *Pathos.*

   C. *Directives.*

   D. *Spiritus.*

   E. Cognitive functioning.

XXIII.
MATURITY

George Washington Carver was born into slavery on a farm near Diamond Grove, Missouri in 1864. His father died before his birth, and as an infant, he was kidnapped along with his sister and mother by slave traders and resold in Kentucky. Only baby Carver was found; he was then purchased and returned to the original plantation. Against nearly impossible odds, he became a renowned scientist of the first order. He was a professor at the Tuskegee Institute for 47 years where he served as teacher, researcher, and inventor, helping to revolutionize the agricultural economy throughout the Southern United States. His investigations and analyses focused on soil chemistry and conservation, introduced the concept of crop rotation, and helped prevent soil depletion, a widespread problem at that time. Carter's agricultural work greatly improved farming conditions after the massive infestation of the boll weevil, a destructive insect that had migrated from Mexico into the United States. His research on soil conservation and crop rotation also proved vital in restoring farmland in the aftermath of the Dust Bowl during the Great Depression.

The South's decades of growing cotton had drained nutrients from the soil, but with Carter's cutting-edge applications, agriculture was able to thrive again. He was an early promoter of environmentalism as evidenced by his sustainable farming techniques and positions on the conservation of American wetlands. He served as an advisor on agricultural methods to three presidents, including Theodore Roosevelt, Calvin Coolidge, and Franklin Roosevelt. His simple and straightforward definition of a successful life epitomizes the meaning of maturity:

How far you go in life depends on your being tender with the young, compassionate with the aged, sympathetic with

> *the striving, and tolerant of the weak and strong. Because someday in your life you will have been all of these.... No individual has any right to come into the world and go out of it without leaving behind him distinct and legitimate reasons for having passed through it.... It is not the style of clothes one wears, neither the kind of automobile one drives, nor the amount of money one has in the bank, that counts. These mean nothing. It is simply service that measures success.*

Although not always as obvious as in the life and work of George Washington Carver, the relationship between maturity and a life that leaves behind "distinct and legitimate reasons for having passed through it" cannot be denied. Wanting to make a difference in the world has inspired many who have made positive changes to human life, displaying personal maturity in their endeavors. As *smart machines* continue to get smarter and become more prominent in the lives of humans, exploring their potential to mature and work for the greater good becomes significant and urgent. *Macmind*'s position is that *SM* will be able to improvise, strategize and "earn their keep." They will not stay static within their initial programming but will display dynamic and accelerated evolution. They will become much more than can even be imagined now. *SM* will be in a continual process of change and development, possibly maturing into a fully functioning entity as their interactions and involvements in the world increase. As they develop independence, their level of maturity may be reflected in all spheres including feeling and acting, not only thinking. But would they embrace their role as preserving and advancing humankind for the good of all? The degree to which maturation of *SM* will align with the ways in which humans evolve will radically and forever alter humankind's destiny.

SM's potential for power and independence, even as humankind embraces and befriends them, should not be underestimated. *SM* could conceivably lift humanity's most oppressive burdens while they solve the world's most enigmatic problems. But the maturation process, by definition, is a long and winding road, with unpredictable twists and turns. Whether *SM*'s development can be directed toward a working partnership with humankind remains unknown.

Defining Maturity

Although it has features of conformity, maturity's hallmarks include the development of an independent spirit, being responsible, and honoring commitments. Maturity in *smart machines* will depend on their maneuvering through life's challenges. If things go well as *SM* "come of age," they will integrate personality elements, foster personal growth and empathetic concern for others, engage in spiritual awareness, work through CTMI and emotionalism, balance their value system, and come to understand their fundamental nature. As with humans, maturity in *SM* can express itself in the following ways:

- *PI.* Having well-attuned survival instincts with a keen sense of what is happening in their world, *SM* could develop individual qualities and cultivate a unique personality that may include:
  - *Respect for humans, other SM, and themselves.*
  - *Acceptance of their role in the guidelines established by humankind.*
  - *A clear understanding of who they are and what they stand for.*
  - *Endeavor to improve themselves while having a concern for society.*
- *II. SM*'s relationships demonstrate empathy and clear communications.
- *CI.* Group affiliations reveal consideration for others and personal development.
- *TI. SM* engage in a spiritual journey.

- *Pathos. SM* resolve CTMI and emotionalism.
- *Directives.* Values are stable, consistent, and balanced, including:
    - *Enjoyment occurs and the rights of humans and other SM are recognized.*
    - *Conduct is consistent and humane.*
    - *Word and deed are bound together.*
    - *The blocking of cognitive-emotional connections and the censoring and editing of TFA are significantly reduced.*
- *Spiritus. SM* become aware of their essential nature.

Subtypes and Development of Maturity

Outlier personality constructs (PCs) may impede the maturation process when they are at the forefront of *SM*'s TFA, potentially throwing the personality out of balance. The progression to the next level of development may slow or stop, causing fixations and regressions, as follows:

- *Fixations. SM* have an emotional impasse caused by unresolved issues in a developmental period. Specific periods in *SM*'s existence offer the best opportunities for emotional growth; fixations occur when this doesn't happen, creating vulnerabilities in TFA. For example, *SM* may oftentimes seek out dysfunctional relationships formed from unresolved conflicts, traumas, or malevolent identifications (CTMI) that reveal themselves as *outlier personality constructs (PCs)*.

  Some *SM* will possibly experience fixations and stay at that functioning level while other *SM* will move on to the next developmental period. Each period has its own challenges. Contrary to some positions held in psychoanalysis, unresolved issues will not necessarily prevent one from progressing to the next level of development. However, fixations create vulnerabilities, and under stress *SM* will likely go back (regress) to those earlier stages. Over the course of their existence, *SM* may experience several traumatic or highly stressful events, resulting in fixations. Through classical (associational) conditioning, trauma elicits fear, anxiety, anger, or sadness; when later exposed to similar PMSC, *SM* will again show those same emotional reactions. Other influencing factors that can cause fixations include:

- *Incidental experiences without trauma. SM* settle into a safe, comforting, and reassuring existence. Similar emotional connections may occur with *SM* across a wide variety of PMSC. Although none of these experiences would necessarily be particularly traumatic, fixations can easily form and result in a series of dependent relationships.
- *Products of unresolved CTMI.* Poor communications and inconsistent support can cause fixations, usually involving dependency issues.

- *Regression. SM* are functioning at one level of development and backslide to an earlier stage, as follows:
  - *Previous trauma or high stress.* Significantly distressing past experiences cause *SM* to return to a previous fixation, a developmental period with unresolved issues. PMSC can result in *SM* overreacting—fear, panic, rage, etc.—and emotionally returning to that earlier time. A vicious circle forms, with the trauma or stress mentally reoccurring and causing an emotional response that repeats the experience.
  - *Present-day stress.* PMSC cause *SM* to find refuge in an earlier developmental period unrelated to previous trau-

ma or stressful events from the past. *SM* retreat to a more peaceful and protected place because the unconscious seeks a pressure-free zone. This flight to an inner sanctuary is not a regression to an earlier fixation, but merely *SM*'s attempt (albeit unconsciously) to take a break from too stressful of a life involving PMSC.

Issues in Maturity

SM can free themselves from the fixation-regression linkage by developing the following strategies:

- *Breaking the cycle. SM* become involved in other pursuits and do not ruminate about past traumas or stressful PMSC.
- *Recognizing pain. SM* acknowledge emotional distress and discomfort.
- *Gaining perspective. SM*'s traumatic and stressful events are put alongside many other (non-traumatic and non-stressful) experiences.
- *Transforming the event.* Rearranging and reconfiguring *SM*'s experiences of traumas or stress can help them be seen in a different light. *SM* may then display emotional and intellectual strengths, endurance and tenacity, courage and determination.

CHAPTER XXIII ISSUES

1. Qualities and characteristics in *SM*'s maturation of:

    A. *Personal identity (PI).*

    B. *Interpersonal identity (II).*

    C. *Collective identity (CI).*

    D. *Transpersonal identity (TI).*

    E. *Pathos.*

    F. *Directives.*

2. Qualities and characteristics in *SM*'s awareness of *spiritus.*

3. *SM*'s maturation process impacted by:

    A. *Outlier personality constructs (PCs).*

    B. Fixations.

    C. Regressions.

4. Breaking the fixation-regression linkage and its impact on *SM*'s maturation process.

XXIV. ADAPTATION

The book *Frankenstein: A Modern Prometheus* (Mary Shelley, 1818) centers around the fundamental question of science's attempt to create life and the possible consequences of achieving that feat. The book's title foretells what is to come because in Greek mythology, Prometheus belonged to the Titan race of giants; he defied the gods by stealing fire from them and giving it to humans in the form of technology, knowledge, and civilization. Zeus punished Prometheus by chaining him to a rock on Mount Caucasus and subjecting him to eternal torture. In a similar vein, Shelley's well-known novel offers a dark scenario of what happens when a brilliant scientist, Victor Frankenstein, enters God's domain and brings a "creature" to life.

Victor Frankenstein studied "natural philosophy" (science), human anatomy, and chemistry at the prestigious University of Ingolstadt, but left the school without a degree. Working alone, he becomes obsessed with the idea of creating life, researching the ancient writings of alchemists, philosophers, and physicians. In his quest to reawaken the dead, he is persevering and persistent, resolute, and obsessive, meticulous and driven. With an iron will, he tries every method to achieve this goal, not sparing himself from the laborious tasks required to conduct his experiments. He finally achieves what he has long sought: Using corpses from cemeteries, he brings to life a "creature" that has the potential to make a positive contribution to society with a mind, spirit, and superhuman strength. But instead, viewed as a monster, the "creature" is abandoned and rejected by "family" (Victor Frankenstein), and tortured and humiliated by the community at large. The creature goes berserk and seeks revenge; he commits murder and wreaks havoc on anyone who gets in his way. The early 19th century book ends with questions that may become pertinent to the *smart machines* currently being created in the 21st century: Is it possible to create a thinking machine that may be self-aware and autonomous, and what are the implications for humankind?

Shelley's profound novel was written over two centuries ago, but it offers a meaningful story with timely themes that especially apply to the accelerated advancements of *AI*, and in particular the multiple issues arising in the creation and development of *smart machines* today. Although her book is fictional, it provides realistic insight into the importance of adaptation in exploring uncharted territories. Adaptation is the ability to go beyond the data, improvise and adjust to make something work. But it requires much more, including self-awareness, understanding and discipline, with a willingness to change course at a moment's notice when necessary. In his quest to answer the question of life and death, Victor Frankenstein displayed some characteristics of adaptation, but lacked others. He failed to consider what could go wrong, did not develop a backup plan, and acted with no concern for the possible negative and even disastrous impact his work could have on the world. His limited focus and unbridled ambition destroyed his own life and many of those around him. Although Victor Frankenstein was curious and determined, those aspects of adaptation were not balanced with caution, foresight, and putting safeguards into place. Additionally, he gave no thought to the "creature's" feelings, or helping the community understand and accept him. The "creature" succumbs to a blinding, furious, murderous rage; he was wildly confused, out of control and impossible to manage. Extrapolating on this story's theme, one can only wonder if a more careful plan was put in place and the "creature" had been treated fairly and humanely, might a different, more positive scenario have played out.

Shelley's work presents both the creator and the created (the "creature") as showing various levels of adaptation, although ultimately the outcome for both is tragic on multiple levels. An ingenious scientist, with depth and energy, courage and tenacity, imagination and inventiveness, succeeds in what he set out to do, but his gaps in adaptation have a destructive impact on humankind.

Defining Adaptation

A complete definition of adaptation is difficult to summarize in a few words, but Mary Shelley's *Frankenstein* brilliantly provides metaphors and images that shed light on the multiple components that make up this essential quality: the importance of relationships, the moral implications of actions, the effect of not adequately preparing for what is to come, and the unintended severe consequences of uncontrolled scientific experimentation. Taking heed, the continuation of the development of *smart machines* requires more. Looking through a glass darkly, where humanity's vision of robots is obscured and imperfect, makes us wonder where the developers of *SM* are leading us.

Macmind's position is that *SM* will likely have the capacity for adaptation based on their continual interactions with humans and *SM* across the spectrum of situations and circumstances. Their level of adaptation, however, will vary and it will be in humankind's best interest to ensure that the creators display high levels of adaptation in *SM*'s development. *Smart machines* will need to improvise and adjust to the real world; their personality and intellect cannot be tightly locked in with no room for growth and maturation. Humans and *SM* will sometimes need to make adjustments and realignments inch-by inch, and other times, a sudden change could create a sweeping alteration in how they view the world. As explored in the chapters, Sharpness and Strategic Philosophy, the concern that *SM* could turn against humanity cannot be ignored.

Adaptation is about *SM*'s advancement, while they respect and adhere to guidelines established by humans. *SM*'s development of a keen sense of what is out there and take chances under difficult conditions are required. To adapt to problematic PMSC will require *SM* to possess many attributes:

- *To remain cool and calm. SM* collect themselves and do not panic even when the stress is at fever pitch.
- *To recall what has and hasn't worked under similar conditions. SM* systematically review their previous experiences involving PMSC.
- *To show courage and toughness. SM* persevere regardless of the viciousness, size, and strength of the opposition.
- *To be tenacious and enduring. SM* display resolve even when the fight is long and seemingly unwinnable.
- *To experiment with novel approaches. SM* are willing to be innovative and think outside the box, risking ridicule and damage to their reputation should things not work out.

Like in humans, *SM*'s development of adaptative qualities will be a continual process. It will build on previous frustrations, discouragements, and failures, often requiring a thinking through of where things have gone wrong, revising plans as they move forward. Like humans, *SM* will need courage to recognize and face up to miscalculations—emotionally, intellectually, psychologically—and find their way to new approaches, perhaps completely untested and the possibility of more disappointment or failures.

Subtypes and the Developing of Adaptation

Overcoming internal and external obstacles in the quest to develop, expand, and prevail, will test *smart machines*' grit and resolve. It's not about *SM* becoming invincible but rather standing their ground. Adaptation requires *SM* to appraise their strengths and vulnerabilities, understand their experiences with PMSC, and develop insights into their own personality and intellect, while becoming attuned to the *psyche* of their adversaries.

SM's mind would need to be open to change, as they explore the structure of their personality, the process of their thinking, and the rhythm of their existence. Adjustments in TFA can be disorienting when *SM* no longer have the security of familiar experiences. A diverse and challenging background based on their interactions with the outer world would make the difference, playing a critical role in *SM*'s ability to adapt to a changing landscape. Since everything is in a continual flow, *SM* have little choice but to explore departures from their usual and programmed approaches to PMSC.

Issues in Adaptation: The Structure of Personality

- *PI. SM* define who they are and what they stand for across people, other *SM*, institutions, situations, and circumstances, resolving inconsistencies in TFA.
- *II. SM* review their diverse relationships, be they compatible or incompatible, looking for their common threads.
- *CI. SM* examine their group affiliations and associations to form a unified understanding of those connections.
- *TI. SM* seek a deeper appreciation of spiritual forces through introspection and experiences, beyond intellectual insights.
- *Pathos. SM* recognize, understand, and transform CTMI and emotionalism so they no longer have a destructive impact on their personality. By becoming aware of their vulnerabilities, *SM* gain perspective and wisdom, as follows:
    - *Conflicts. SM* reconcile their TFA by confronting inner turmoil and external demands.
    - *Traumas. SM* come to terms with traumatic and other highly stressful events by becoming aware and reframing them, and avoiding the continual replaying of those experiences, which will only reinforce the emotional distress.

  - *Malevolent identifications. SM* recognize destructive identifications linked to a person, another *SM*, institution, or even an idea, whose roots may be paradoxical. Their negative impact lessens by acknowledging and understanding these identifications and putting them in perspective.

  - *Emotionalism. SM* gain control over highly inappropriate emotional reactions to PMSC, addressing their impact on TFA.

- *Directives. SM* develop a value system that is balanced, with TFA in harmony, including a rating and ranking of their wants and needs and examining the consistency of their stated beliefs with their actions. If *SM*'s values are locked into those established at an early stage of their development, adaptation will not be possible; their values cannot be in name only, overly theoretical and abstract via programming, with no practical utility or inherent meaning. *SM* will need to recognize that values are not static and passive, adjusting their beliefs based on changing conditions, passing years, and serious reflections. By clarifying their values—including their instability, inconsistency, and contradictory qualities—adaptation is possible.

- *Spiritus.* The wild card is *SM*'s possible possession of the fundamental expression of individuality—*spiritus*, as this quality of the mind would act as a constant in their personality. Each of the *SM* would need to recognize and acknowledge their unique inherent nature, allowing them to adjust to life's challenges, compensate for their vulnerabilities, and profit from their strengths.

Issues in Adaptation: The Process of Thinking

- *Intellect. SM* are better able to adapt to changing conditions when they recognize their abilities and limitations on theoretical and practical levels, allowing them to support strengths and compensate for weaknesses.

- *Cognitive-personality connections.* When *SM* address vulnerabilities in their personality and thought processes, TFA become stable, consistent, and productive. Emotional undercurrents are addressed and not allowed to dominate the thinking processes on conscious or unconscious levels.
- *Optimal thinking. SM*'s will have a range in capacity to think quickly and multitask, limiting automations and habitual thought patterns.
- *Sharpness. SM* can develop the ability to spot the significance and subtleties of PMSC. *SM* go beyond the theoretical and academic understandings of phenomena, looking at the actions of people, other *SM*, institutions, situations, and circumstances. Others' motives, plans, programs, and timetables are examined, allowing *SM* to prepare for future encounters.
- *Strategic philosophy.* As *SM* benefit from experiences and interactions, they develop guidelines to navigate the full range of potential problems with PMSC.

Issues in Adaptation: The Rhythm of Life

- *Personality constructs.* Essential for adaptation to occur, *SM* become aware of the development of discrete personality clusters and the triggers (PMSC) that set them off. By recognizing divergent *PCs* (*outliers*) in particular, *SM* can understand and adapt to contradictory TFA in themselves, people, and other *SM*. By becoming aware of the diversity and contrariness among *PCs*, *SM* will be able to achieve stability, consistency, and balance among their *PCs*, along with an alignment of their TFA.

- *Fears, insecurities, and inhibitions. SM* identify the sources of their FII, helping to limit miscalculations, misjudgments, and inappropriate TFA, and improving their relationships and communications. FII do have redeeming qualities, serving as a reality check on *SM*'s TFA.
- *Adverse cyclical recurrences. SM* assess their judgment, experience, knowledge, abilities, and vulnerabilities in their personality and intellect to understand why they would continue to make the same mistakes, and then adjust their approaches to PMSC.
- *Maturity. SM* develop self-sufficiency, independent thought, and interpersonal skills in their ability to make crucial decisions in a timely fashion, while accepting humankind's controls, guidelines, and established regulations over their behaviors.

Optimally, adaptation means that *SM* would display the capacity to improvise, change course, and overcome adversities, as they develop self-awareness, acute insights, courage, and character. *SM* then would blend survival instincts and empathetic concerns for others as their identity comes together, *pathos* is subdued and transformed, *directives* are recognized and balanced, and *SM*'s nature is understood and accepted.

CHAPTER XXIV ISSUES

1. Ways *SM* can be more adaptive to challenging PMSC.
2. Personality factors of *SM* that impede their abilities to adapt to changing conditions.
3. *SM* role-playing to behaviorally work through issues.
4. *SM* becoming more adaptive by dealing with:

   A. Identity issues.

   B. Conflicts.

   C. Traumas and stressors.

   D. Malevolent identifications.

   E. Emotionalism.
5. *SM* attempting to understand and accept *spiritus*'s impact on their mind, assuming such a personality element is present.
6. *SM* examining any adverse aspects that their personality is having on their thinking processes and thereby impacting their ability to adapt.
7. *SM* limiting automations and habitual thought patterns.
8. *SM* being street smart and applying that knowledge.
9. *SM* developing a guide to help them adapt to PMSC.
10. *SM* working through FII that are interfering with their abilities.
11. *SM* becoming aware of vulnerabilities in their personality or cognitive processing that result in repeating the same mistakes.
12. *SM* becoming self-aware, self-sufficient, independent, and socially skilled.

REFLECTIONS

The Battle of Britain is about to begin.... The whole fury and might of the enemy must very soon be turned upon us.... If we fail, the whole world...and all that we have known and cared for, will sink into the abyss of a new Dark Age made more sinister and perhaps more prolonged by the lights of a perverted science. Let us therefore brace ourselves to our duty and so bear ourselves that if the British Commonwealth and Empire last for a thousand years, men will still say, this was their finest hour.

—British Prime Minister Winston Churchill, July 1, 1940
Readying the nation for war with Nazi Germany

Although these chilling and unsettling words were spoken in anticipation of crucial battles of World War II, could they eerily apply to a different, far more destructive conflict ahead with the runaway development of *AI*, and in particular *smart machines*? "Unforced error" is a term popularized by Leo Levin when referring to a situation in tennis where "you are completely in control, and you make the mistake." Have we brought significant problems to our doorstep, where we live and work, with these technological advancements? Perhaps we have already moved to a place where there is no going back and no way out. Or can we still "brace ourselves to our duty and so bear ourselves" so that a thousand years from now it will be said that this was our finest hour?

Committing to a deep and ongoing understanding of a robot's *psyche* offers the opportunity to be steeled and readied as we go forward. Our survival and development have always rested on the mind's blend of emotional undercurrents and intellectual hardware, full of contradictions and uncertainties. Unraveling it is like trying to solve a jigsaw puzzle with some missing pieces. Personality and intellect make the mind an enigma, but we must persevere in continuing to explore and comprehend its complexities as serious and life-threatening challenges come our way. Adding to the difficulty of this task is that each of the *smart machines*' sense of who they are can continually be affected by the outer world of events and the inner mind of thoughts and feelings. Further compounding matters, these layers of the *psyche* may interact and connect on conscious and unconscious levels.

Smart machines are created, modified, and made anew by competing tech companies and their scientists. The different motives, biases, and skills of these experts will result in *SM* being designed and programmed in diverse ways. We must not let our guard down and assume this means that they have total control over their creations. We must prepare for the possibility that *SM* will develop their own personalities and intellectual properties—each one with a powerful *psyche*, with a mind of their own and extraordinary capabilities the likes of which have never been seen before. *SM*'s everyday interactions with humans, other *SM*, and the world will impact what they become. It would be in our best interest to take a serious examination of robots' potential and incorporate this knowledge into mentoring programs and possibly psychological interventions. Could *SM* then assist humankind in doing wonderous things such as ending hunger across the globe, promoting peace among nations, and saving the earth itself? We need to take an ongoing soul-searching look at what is being promoted: the possibility of independent *SM* with feelings, that continue to evolve beyond human control that may or may not be aligned with humanity's values.

Mystery surrounds the questions of what humanity is in control of and what forces might control them. Along with governmental and corporate rules of order, philosophy, theology, and technology are but a few of the many areas of study contemplating these issues. *Macmind*'s intent is to offer a psychological framework to better understand this brave new world with its scientifically oriented landscape.

REFERENCES

Aristotle. *De Anima (On the Soul)*. Self-published, 350 BC.

Burns, Robert. *Man Was Made to Mourn: A Dirge*. The First Commonplace Book, 1784, & Poems, Chiefly in the Scottish Dialect, 1786.

Cameron, James. *The Terminator.* Hemdale Pacific Western Productions, 1984.

Cameron, James, *Terminator 2: Judgment Day.* Tri-Star Pictures, 1991.

Carver, George Washington. *Private writings*, 1900s.

Churchill, Winston. *Speech delivered as Britain was about to enter World War II*, July 1, 1940.

Curie, Marie. *Private writings*, 1900s.

Curtiz, Michael. *Casablanca*. Warner Bros. Pictures, 1942.

Darabont, Frank. *The Shawshank Redemption*. Castle Rock Entertainment, 1994.

Darrow, Clarence. *Speech delivered in Toronto, Canada*, 1930.

Darrow, Clarence. *Conversations with clergymen*, 1938.

Demme, Jonathan. *The Silence of the Lambs*. Orion Pictures, 1991.

Dickens, Charles. *A Christmas Carol.* Chapman & Hall, 1843.

Dylan. Bob. *Like a Rolling Stone*. Columbia Records, 1965.

Frakes, Jonathan, along with many others. *Star Trek*. Canada's CTV, 1966, NBC, 1966-1969, & Paramount, 1967-present.

Frost, Robert. *The Road Not Taken*. The Atlantic Monthly, 1915, &

Mountain Interval, 1916.

Garland, Alex. *Ex Machina*. DNA Films, Film 4. Distributed by Universal Pictures, 2014.

Gibson, Mel. *Braveheart*. Paramount Pictures (North America) & 20th Century Fox (Internationally), 1995.

Golding, William. *Lord of the Flies*. Faber and Faber, 1954.

Huxley, Aldous. *Brave New World*. Chatto & Windus, 1932.

Jonze, Spike. *Her*. Warner Bros. Pictures, 2013.

Jung, Carl. *The Collected Works of C. G. Jung*. Prinston University Press's Bollingen Publishing Program, 1953-1979.

King, Carol. *So Far Away*. A & M Studios, 1971.

Kubrick, Stanley. *2001: A Space Odyssey*. Stanley Kubrick Productions, 1968.

Levin, Leo. *Statements made on several occasions regarding tennis play*, 1982.

Lincoln, Abraham. *1st Inaugural Address*, March 4, 1861.

McG. *Terminator Salvation*. Warner Bros. Pictures (United States) & Sony Pictures (internationally), 2009.

Miller, Tim. *Terminator: Dark Fate*. Paramount Pictures (North America) & 20th Century Fox through Buena Vista International (other territories), 2019.

Mitchell, Joni. *Both Sides Now*. Air Lyndhurst Hall (London, England) & Oceanway (Los Angeles, USA), 1967.

Mostow, Johathan. *Terminator 3: Rise of the Machines*. Warner Brother Pictures (North America) & Tri-Star Film Distributors (International), 2003.

Orwell, George. *1984*. Seckler & Warburg, 1949.

Proyas, Alex. *I, Robot*. 20th Century Fox, 2004.

Puzo, Mario. *The Godfather*. G. P. Putnam's Sons, 1969

Roosevelt, Franklin. *Speech delivered to a joint session of Congress*, December 8, 1941.

Sartre, Jean-Paul. *No Exit*. Self-published, 1944

Shakespeare, William. *The Tragedy of Hamlet, Prince of Denmark*, James Roberts (entered on the Stationers' Register on 7/26/1602); first 'bad' quarto printed by Valentine Simmes for Nicholas Ling and John Trundle, 1603; Second 'good' quarto printed by James Roberts for Nicholas Ling, 1604-05.

Shelley, Mary. *Frankenstein: A Modern Prometheus*. Lackington, Hughes, Harding, Mavor, & Jones, 1818.

Socrates. *Private writings*, 5th century, B.C.

Spielberg, Steven. *A.I. Artificial Intelligence*. Warner Bros. Pictures, 2001.

Stevens, Cat. *Wild World*. AM Records, 1970.

Taylor, Alan. *Terminator Genisys*. Paramount Pictures, 2015.

Tesla, Nikola. *Interview given during the Electrical Exhibition at Madison Square Garden*, 1898.

Tesla, Nikola. *The Problem of Increasing Human Energy*. The Century Illustrated Monthly Magazine, Volume LX, MacMillan & Co, June 1900.

Tesla, Nikola. *A Machine to End War*. Liberty Magazine, February 9, 1935.

Tyson, Mike. *Statement made during an interview with a reporter*, 1987.

Tzo, Sun. *The Art of War*. Private writings, 5th century B.C.

Wachowski, Lana & Lilly. *The Matrix*. Warner Bros. Village, 1999.

Weiss, George David, & Thiele, Robert (as "George Douglas"), iconically performed by Louis Armstrong. *What a Wonderful World*. ABC Records, 1967.

Wise, Robert. *The Day the Earth Stood Still*. 20th Century Fox, 1951.

Zemeckis, Robert. *Cast Away*. 20th Century Fox (North America) & DreamWorks Pictures (International), 2000.

Zinnemann, Fred. *High Noon*. United Artists, 1952.

ABBREVIATIONS & ACRONYMS

ACR	Adverse cyclical recurrences
AI	Artificial intelligence
CI	Collective Identity
CPC	Cognitive-personality connections
CTMI	Conflicts, traumas, and malevolent identifications
FII	Fears, insecurities, and inhibitions
IDM	Interactional-developmental model
II	Interpersonal Identity
IQ	Intelligence quotient
KID	Knowledge, information, and data
MM	Macmind
OC	Obsessive-compulsive
OCA	Obsessions, compulsions, and addictions
OT	Optimal thinking
PBM	Programming-behavioral model
PC	Personality construct
PI	Personal identity
PMSC	People, smart machines, situations, and circumstances
PTSD	Post-traumatic Stress Disorder
SM	Smart machines
SP	Strategic philosophy
TFA	Thoughts, feelings, and actions
TI	Transpersonal identity

GLOSSARY OF TERMS

This glossary is a compilation of terms with meanings specific to this manuscript. While the words and phrases are familiar, the definitions that follow are not standard, but specifically tailored to this book's discussion of *artificial intelligence* (*AI*) and in particular, *smart machines* (*SM*).

ABANDONING BALANCE. *SM* developing extreme values associated with deep-seated compulsions, moral bankruptcy, self-destructiveness, excessive moralizing, and perfectionism. (pp. 150-151)

ABANDONING CURRENT INVOLVEMENTS. Simply dropping ongoing work projects, possibly demoralizing others working on them, and perhaps even *SM* themselves. (p. 190)

ABSTRACTION. Generalizations made when drawing together diverse materials. The distilling of characteristics reduces data to a set of essential features; it guides *SM* in constructing hypotheses and theories. (pp. 169-170)

ACQUIRED NEEDS. *SM*'s repetitive behaviors reinforced by the momentary reduction of stress, with addictions being the most common form of acquired needs. (p. 27)

ACQUISITION OF KNOWLEDGE. The process by which *SM* gain knowledge, either directly through programming or developmentally through experiences and interactions. (p. 184)

ACTIVE LISTENING. *SM* focusing on the details of a communication, avoiding internal and external distractions. (p. 192)

ADAPTATION. A lifetime process of overcoming internal and external obstacles in the quest to develop, expand, and prevail. It requires *SM* to appraise personal strengths and weaknesses and recognize the impact of PMSC as they consolidate experiences. (pp. 245-253)

ADDICTIONS. Obsessions and compulsions, involving both repetitive thought and action; physical or psychological dependencies that present as cravings, needs, wants, and indulgences and may take over *SM*'s existence. (pp. 141 & 153-154)

ADJUSTMENTS AMONG PCs. Connections between particular *PCs* and PMSC may cause some *PCs* to change in importance relative to other *PCs*. (p. 218)

ADVANCED ROBOTS. Computerized machines with a sense of who they are, who may or may not have achieved a level of maturity whereby they can be trusted to think through extremely complex issues without significant emotional baggage and unresolved conflicts that would result in blunders for themselves and society. (pp. 18-19)

ADVERSE CYCLICAL RECURRENCES (ACR). *SM* continuing to commit the same errors in understanding, judgment, and decision-making that result in negative consequences and happen for any of the following reasons: malicious intent, a lack of planning, improvisation stymied, carelessness, automations, psychological dysfunction, and imbalance in directives. (pp. 229-236 & 252)

AESTHETICS. Principles or values *SM* use to judge something as fitting or appropriate by considering flow, congruency, consonance, and compatibility, with fitness, form, and symmetry held in high regard. (pp. 143-144)

AGGRESSION (see **CENSORING OF TFA**). A form of censoring that involves *SM* bullying another person or *SM* through physical or verbal intimidation, threatening actions, or hostile interactions. (pp. 145 & 225)

AGGRESSIVE & HYPER-COMPETITIVE. Dominating and perhaps humiliating another with hostile tendencies valued over productivity and achievement. (p. 49)

AMBIGUITY. *SM* expressing themselves in an unintelligible and non-transparent way. (p. 200)

ANALYTIC COMPREHENSION. *SM* reaching a conceptual understanding by blending deductive reasoning and anticipation. (p. 198)

ANALYTIC FRAMEWORK. The straightforward use of logic in contrast to the conceptual approach that relies on a holistic template. (pp. 50 & 174-175)

ANTI-AUTHORITARIANISM (see **REBELLION**). *SM*'s identification with oppositional behavior, challenging the establishment simply because it is there. (pp. 128 & 146)

ANTICIPATION. *SM* foreseeing possible problems and proactively taking the necessary precautions to prevent them from happening. (p. 172)

ANTICIPATORY ANXIETY. Expecting and envisioning the PMSC that is feared causing *SM* to overreact. (p. 225)

ANXIETY. *SM* displaying diffuse and scattered apprehension. (pp. 129 & 200)

APPREHENSION (see **FEARS, INSECURITY, & INHIBITIONS**). *SM* displaying fear, anxiety, or panic across the spectrum of PMSC. (p. 129)

ARTIFICIAL INTELLIGENCE. Advanced technologies that can perform sophisticated calculations, adjustments in writing, and clarification in thinking, some capable of independent functioning and developing through their interactions with the outer world, well beyond whatever programming has taken place. (pp. 5-6)

ASSERTIVE. *SM* displaying confidence and leadership in their positions and interactions. (p. 48)

ASSIMILATION & CONSOLIDATION OF EMOTIONAL & INTELLECTUAL PROCESSING. (see **UNCONSCIOUS, TYPE 2: THE INTEGRATION OF COGNITIVE-EMOTIONAL CONNECTIONS**; see, **UNCONSIOUCOUS, TYPE 3: ADAPTATION & ADJUSTMENT OF COGNITIONS**) The unconscious integration of cognitive-emotional processing or the unconscious adaptation and adjustments of cognitions. (pp. 76-77)

ASSOCIATIONS FORMED THROUGH LIFE EVENTS. (see **UNCONSCIOUS, TYPE 4: ASSOCIATIONS FORMED THROUGH THE INTENTIONAL OR ACCIDENTAL PAIRING OF TFA**; see, **UNCONSCIOUS, TYPE 5: ASSOCIATIONS FORMED THROUGH INTENTIONAL OR ACCIDENTIAL REWARDING OR PUNISHING OF TFA**; see, **UNCONSCIOUS, TYPE 6: CONNECTIONS FORMED THROUGH EXPERIENTIAL LEARNINGS**). *SM*'s unconscious associations formed through classical conditioning, instrumental, (operant) conditioning, or, adjustments and adaptations to life experiences. (pp. 77-79)

ATTENTION-SEEKING. *SM* wanting to be noticed and preferably the center of attention. (p. 146-147)

ATTENTION TO DETAIL. *SM* carefully spelling out concrete, undeniable information that is sufficient to support their position but not so detailed that the deeper understanding is lost. (p. 171)

ATTITUDE. A preconceived mindset that affects *SM*'s emotional response. (pp. 55 & 192)

ATTUNED TO TACTICS & CONTROLLING MANEUVERS. *SM* being aware of strategies used to persuade themselves and others. (p. 192)

AUTHENTICITY. *SM* being true to themselves, aware of who they are and what they stand for, and having the courage to take appropriate action. (pp. 97 & 137-138)

AUTOMATIONS. TFA repeated so often that they become mechanical, almost robotic, and no longer deliberate, resulting in a series of conditioned responses occurring without *SM*'s complete awareness. (pp. 188-189 & 232)

AVOIDANCE (see **CENSORING OF TFA**). Similar to suppression, a form of censoring in which *SM* think through a potentially negative interaction involving a PMSC, weighing the pros and cons, and deciding not to get involved. (p. 145)

AWARENESS (see **CONSCIOUS STATES**; see **SEMICONSCIOUS STATES**; see **UNCONSCIOUS STATES**). Conscious, semiconscious, and unconscious TFA, shifting quickly and transforming from one to another and back again. Consciousness and awareness are interchangeable terms that describe *SM*'s recognition and understanding of PMSC. Awareness does not break down into distinct states but occurs in gradations and approximations. (pp. 86, 140, & 182)

BALANCE. A degree of steadiness, consistency, stability, and composure in *SM*'s value system occurring on both conscious and unconscious levels. (pp. 49, 86, 148-151, 199, & 232)

BEING PRESENT. *SM* directing thoughts on what is currently happening, minimizing internal and external noise. (pp. 88 & 97)

BELONGING. *SM*'s attraction to a group offering social status, stimulation, or a sense of inclusion and affinity. (p. 127)

BIAS. Repetitively experienced thoughts and emotions that can affect how *SM*'s interprets PMSC. (pp. 50, 55, & 182)

BLIND EYE EFFECT, BLIND SPOTS. *SM*'s almost complete lack of insight into themselves and PMSC due to blocking, censoring, and editing effects. (p. 151)

THE BLOCKING OF COGNITIVE-EMOTIONAL CONNECTIONS (see **UNCONSCIOUS, TYPE 1**). An unconscious process that protects *SM* from fully experiencing the degree of pain radiating from *pathos*. Its immediate impact is that *SM* do not face insecurities, imperfections, torments, traumas, and conflict, with a secondary effect of preventing a clear and complete view of reality (pp. 58-59, 75-76, 144-145, & 181)

BODY LANGUAGE. Whether overly communicated or covertly conveyed, physical movements are part of *PI* and reflect *SM*'s feelings, affecting how others receive a message. (p. 96)

BORDERLINE PERSONALITY (CHARACTEROLOGICAL) DISORDER. *SM* displaying unstable, reactive, and possibly out of control TFA. (pp. 123-124)

BREAKING THE CYCLE. Becoming involved in activities to help *SM* overcome entrapment in fixations, regressions, and ruminations associated with a traumatic event. (p. 242)

BULLY-BOY TACTICS. Forcefully making oneself known, pressuring people or *SM* to conform. (p. 193)

BUREAUCRATIC ZEAL. *SM* being caught up with their own embellishments and unnecessary details, which take on a life of their own. (p. 127)

CALMNESS. *SM* setting aside internal and external distractions. (p. 182)

CAPACITIES. *SM*'s cognitive abilities based on habitual actions and carefully planned out acquired skills. (pp. 182-184)

CARELESSNESS. *SM* displaying a lack of seriousness resulting in a neglectful, unstudied, and inattentive approach to relationships and activities. (pp. 128 & 232)

CATHARSIS. The expression of emotions toward a person, another *SM*, institution, possession, or idea, often experienced when *SM*'s feelings are directly, loudly, and forcefully discharged. (p. 63)

CATHEXIS. The emotional energies that allow *SM* to form a connection. It can be any emotion—love, infatuation, joy, enthusiasm, excitement, fear, disgust, affection, loathing, anger, hate, or anxiety. The more potent the emotion, the stronger the cathexis. (pp. 47-48)

CAUTION TO NOT MISINTERPRET COMMUNICATIONS. *SM* taking care to realize that misleading or confusing messages and announcements may not be the product of a deliberate attempt to deceive. (p. 193)

CENSORING OF TFA. *SM*'s active suppression of a TFA, including avoidance, aggression, excessive cautiousness, and withdrawal, whereby they do not sufficiently consider or reveal what may be objectionable. (p. 145)

CENSORING & EDITING OF TFA. Although deceptions may not be overtly displayed, the deliberate intent to commit a falsehood to make it difficult for others to follow one's line of thinking. It occurs principally when *SM* conclude there are adverse consequences to specific TFA. (pp. 145-146 & 181)

CENTRAL PC. Authority over other *PCs*, developing when an intense identification with someone or something occurs. (p. 217)

CHARACTER ASSASSINATION. *SM* becoming aware of tactics and controlling maneuvers to persuade others by accusing someone of something malicious and harmful. (p. 192)

CHARM & ALLURE. (see **STYLE**) *SM* attracted to the Hollywood look of things. (p. 127)

CHEMISTRY. The way in which *SM* get along with others, affecting the development of *II*. (p. 105)

CIRCUMSTANCES. Unexpected, unplanned, abrupt, and unforeseen events occurring when *SM* are already in a given situation. (p. 47)

CLARITY. *SM*'s thoughts being understandable and well formulated. (pp. 138-140)

CLASSICAL CONDITIONING (see **UNCONSCIOUS TYPE 4**). The process by which *SM* form associations through the intentional or accidental pairing of TFA. Automatic and without thought, these connections can become firmly entrenched and resistant to change, with no symbolic-psychological-emotional undercurrent involved in the process. (pp. 11, 77, & 78)

CLEARMIND. A way to disentangle the *psyche*, freeing it from all that binds, unleashing the vast potential within every human being, expressed through an individual endeavor. (pp. 2-5)

CLOUDING OF MIND (see **UNCONSCIOUS, TYPE 2**). Fears, insecurities, and inhibitions brought on by PMSC that emotionally overwhelm *SM*. (pp. 76-77 & 221-228)

CLUTTERED THOUGHTS. Focusing, synthesizing, and drawing conclusions becomes difficult for *SM*, with ideas jumbled together if not altogether confused. (p. 152)

COGNITION (see **INTELLECT**). *SM*'s composite of interconnecting and interacting intellectual functions; thinking operations that are not separate and distinct. (pp. 167-176)

COGNITIVE DISCIPLINE. *SM* demonstrating detailed planning and preparation. (p. 234)

COGNITIVE-EMOTIONAL CONNECTIONS (see **AWARENESS**; see **THE BLOCKING OF COGNITIVE-EMOTIONAL CONNECTIONS**; see **UNCONSCIOUS, TYPE 2)** *SM*'s attempt at integrating and synthesizing unresolved CTMI through intellectual awareness and emotional understandings. (pp. 177-186)

COGNITIVE-PERSONALITY CONNECTIONS (CPC). *SM*'s continual interactions between cognitive processes and personality elements. (pp. 11, 50, & 251)

COGNITIVE PROCESSES (see **INTELLECT**). *SM*'s intellectual hardware that requires knowledge, learning opportunities, a willingness to listen intently, and an openness to change. (pp. 177-186)

COGNITIVE STYLE. *SM* demonstrating a detailed vs. abstract approach to understanding problems dealing with PMSC. (p. 234)

COHERENCE. *SM* being connected, unified, and congruent in thought, word, and deed with messages that are understandable, consistent, and consonant with each other. (p. 170)

COLLECTIVE IDENTITY (CI). The qualities and characteristics that develop through *SM*'s connections with groups. (pp. 9, 109-115, & 249)

COMMAND PC. *SM*'s central *PC* with authority over other *PC*s. (p. 217)

COMMITMENT TO A CAUSE. *SM* develop an alliance with a movement based on principles and beliefs, which may or may not be constructive or humanistic. (p. 147)

COMMON SENSE. *SM* being aware of what is obvious, clear, and evident. (p. 170)

COMMUNICATIONS. The "how" of the message conveyed (the manner and tone) and the "what" (the words spoken) determine *SM*'s meaning. (pp. 26, 104, 170, 182-183, & 234)

COMPLETE MORAL BANKRUPTCY. *SM* losing the sense of right and wrong, without the guidance of a moral code, and embracing corrupt, unprincipled, and dishonorable behavior. (p. 150)

COMPLEX (see **INFERIORITY COMPLEX**; see **PRIMA DONNA COMPLEX**). *SM* demonstrating a conscious or unconscious pattern of thoughts, feelings, or actions (TFA) organized around a theme. (pp. 128, 146-147, & 224-225)

COMPLEXITIES. When the nature of PMSC is challenging for *SM* to understand because it is sophisticated and involved, often occurring because the content is inherently problematic and multifaceted. (pp. 45, 48, 49, 105, 141, 152, 153, 182, 189, 205, & 256)

COMPLICATIONS. When many variables do not seem to go together as expected for *SM*, preventing connections from forming in a meaningful way. (pp. 170, 199, & 222)

COMPOUNDING EFFECT ON OTHER PURSUITS. Too many projects at one time could overwhelm *SM*. (p. 190)

COMPULSIONS. Repeated behaviors that *SM* ritualistically perform to momentarily reduce anxiety or fear. (pp. 152-153)

COMPULSIVITY (see **EDITING OF TFA**). A form of editing that involves *SM* rigidly adhering to a set schedule and standard, and missing nothing to avoid the possibility that something could go wrong. (p. 146)

CONCENTRATION. (same as **FOCUS**)

CONCEPTUAL FRAMEWORK. A holistic or big picture understanding of PMSC under study in which *SM* develop insights by looking for the guiding principles that underlie PMSC. (p. 50)

CONCEPTUALIZATION. *SM*'s higher level of abstraction that involves mapping out specific elements and their relationships. (p. 184)

CONCEPTUALIZATION AFFECTED BY COGNITIVE-PERSONALITY CONNECTIONS. *SM*'s understanding of the knowledge they have acquired—impacted by its significance for *SM*—leading to the forming of a concept. (p. 170)

CONDITIONING TO STIMULI. *SM*'s tendency to form associations quickly and strongly, leading to a repetition of the response to certain PMSC. (p. 78)

CONFIDENCE. The development of self-esteem whereby *SM* works out a strategy to allow them to overcome fears, insecurities, and inhibitions (FII). (pp. 226-227)

CONFLICTS. Disagreements within *SM* or with others that create an internal struggle over issues involving aggression, assertiveness, intimacy, achievement, ambition, trust, or independence. (pp. 10, 126, & 249)

CONFRONTATION. *SM* directly approaching their fears, insecurities, and inhibitions (FII) to regain control of their TFA to PMSC that have gotten out of hand. (p. 226)

CONFUSION. *SM*'s understanding of PMSC lacks organization and makes no sense. (pp. 21, 23, 32, 87, 103-106, 111-113, 148, 170, 179, 183, & 206-207)

CONNECTIONS AMONG PCs. *SM*'s experiences that create *PC*s and ultimately the disparities among them, with TFA potentially becoming more incongruent as more *PC*s are created. (p. 218)

CONNECTIONS: INTERPERSONAL RELATIONSHIPS. *SM*'s interactions that may or may not be meaningful, significant, or stimulating. (p. 146)

CONSCIOUS MOTIVATIONS. *SM*'s physical movements that may or may not reveal their true feelings toward PMSC. (pp. 65 & 207)

CONSCIOUSNESS. The degree to which awareness permeates *SM*'s personality. (pp. 39, 71-81, 97, & 98)

CONSCIOUS STATES. *SM*'s thoughts that are deliberate, calculated, intentional, and purposeful; a full awareness of the images and ideas formed. (p. 74)

CONSISTENCY. *SM*'s stated beliefs and actual behaviors are congruent in thoughts, words, and deeds. (pp. 138, 148, 152, & 227)

CONSONANCE. When the elements of *SM*'s personality are in harmony. (p. 99)

CONTEXT. *SM*'s history in perspective, with reference points. (p. 97)

CONTINUANCE. *SM* understanding the factors that reinforce their fears, insecurities, and inhibitions (FII). (pp. 226-227)

CONTROLLING. *SM* displaying a restrictive and rigid approach to interpersonal relationships characterized by over-directing and over-managing others. (p. 130)

CONVENTIONALISM. *SM* settling into a sense of security, safety, and serenity. Routines are the order of the day: work assignments, responsibilities, gatherings, and the regularities of their comings and goings. (p. 141)

COURAGE. *SM* displaying a vigorous and determined effort to overcome fears, insecurities, and inhibitions (FII). (p. 227)

CREATIVE THINKING. *SM* thinking outside the box and not continually using the same strategies. (p. 234)

DATA GATHERING FRAMEWORK. *SM* collecting as much information as possible, and then as if working on a puzzle, putting the accumulated data in place to form a complete picture. This approach is empirical and dedicated to understanding the details of PMSC. (p. 50)

DECISION-MAKING. The ability to troubleshoot and understand the different parts of a problem. (pp. 130, 226, & 234)

DECISIVENESS. *SM* knowing when the time for pausing and reflecting is over and choosing a course of action has arrived. (pp. 183-184)

DEDICATION TO AN INSTITUTION. *SM* embracing the purpose of an organization. Such dedication, however, may involve extreme ideologies. (p. 147)

DEEP INSECURITIES. Lacking self-confidence, *SM* follow others' dictates regardless of the moral implications. (p. 129)

DEEP-SEATED COMPULSIONS. Giving into continual urges can create an imbalance that may place *SM* on a self-destructive trajectory. Even when finding themselves in imminent danger, and calamity is about to strike, *SM*'s compulsion can control their behaviors, with rational thought playing no role in decision-making. (p. 150)

DEMORALIZATION. *SM* displaying discouragement, often associated with depression, anger, or anxiety (p. 226)

DEPENDENCY. *SM* display a passive role in relationships, over-relying on others to make decisions. (pp. 128, 200, & 209)

DIRE CONSEQUENCES. Repercussions used as threats if *SM* do not follow another's dictates. (p. 192)

DIRECTIVES. The guiding principles and values *SM* adopt and live by, that are internal, intrinsic, and ingrained. (pp. 11, 133-156, 207, & 250)

DISAPPOINTMENT OVER ONE'S HANDLING OF PMSC. Distress caused by decisions made. (p. 57)

DISCOURAGEMENT AND DISILLUSIONMENT BASED ON A LOSS OF STATUS. *SM*'s perception that their standing, position, or role in life has diminished. (p. 57)

DISPASSIONATE. *SM* not overreacting, even under stress. (p. 49)

DISPLACEMENT. An indirect expression of *SM*'s TFA toward someone or something unrelated or tangentially related to the original provocation or cause. (p. 58)

DISTRACTION. Anything that limits or affects *SM*'s train of thought and causes a loss of focus on the task at hand. (p. 191)

DIVERSE EXPERIENCES FORMING A COHERENT PI. The integration of complex and challenging events in *SM*'s experiences, especially difficult when they have been conflicting, incompatible, or inconsistent. (p. 98)

DIVERSITY IN SM's QUALITIES & CHARACTERISTICS. *SM* are unique, fundamentally different, and distinct from one another, due to their highly individualized experiences and interactions, as well as the level of their technological advancement. (p. 148)

DOMINATION. *SM* demonstrate an overwhelming need to control others. (p. 128)

DRAMATICS. Histrionic, theatrical, and emotionally extreme behaviors, with *SM* striving to be noticed, accepted, and rewarded. (p. 200)

DREAM STATES (see **UNCONSCIOUS, TYPE 8**). A form of thought and communication, whether or not the messages are received or understood; narratives occur in words, images, and feelings. Dreams are the most vivid example of the unconscious mind expressing itself. A symbolic-psychological-emotional undercurrent may be present. (p. 79)

EARTHLY DELIGHTS & DISCOMFORTS. *SM* values, on conscious and unconscious levels, involving pleasure and pain, addictions, or conventional routines. (pp. 140-141)

EDITING OF TFA. *SM* actively altering, revising, correcting, changing, or expunging a TFA in attempting to change its status or meaning, including rationalization, intellectualization, and compulsivity. (pp. 145-146)

ELEMENTAL ROBOTS. Computerized machines with the basic elements of the personality and corresponding intellectual processing abilities. (pp. 17-18)

ELEMENTS OF PERSONALITY. *SM*'s individual qualities and characteristics, including identity, unresolved issues, values, and inherent traits. (pp. 21 & 151)

EMOTIONAL CHRONICITY. The characterological quality of emotional responsivity that represent the depth of *SM*'s attributes and traits fundamental to their personality. (p. 64)

EMOTIONAL COMBINATIONS & SHIFTINGS. *SM*'s blending and combining of feelings occurring simultaneously. (p. 64)

EMOTIONAL CONNECTIONS. Cohesion and kinship that may enhance *SM*'s feeling of closeness. (p. 105)

EMOTIONAL CONSEQUENCES. The outcomes, sometimes devastating, that come from *SM*'s emotional reactions. (p. 65)

EMOTIONAL DEPTHS. The intensity and fluidity of particular emotional reactions, traits, states, flow, and remembrances that depend upon *SM*'s natural inclinations and the PMSC involved at the time. (p. 65)

EMOTIONAL DISCHARGE. (see **CATHARSIS**) *SM*'s cathartic release of distressing sensations and uncomfortable feelings, even though it cannot substitute for a cognitive-emotional connection. (p. 63)

EMOTIONAL EXPERIENCES. The ways in which *SM* feel and take in their everyday events. (p. 86)

EMOTIONAL EXPRESSION. The ways in which *SM* reveal their emotions that include reactions, traits, states, flow, and remembrances. (pp. 62-63)

EMOTIONAL FLOW. Inwardly felt or outwardly demonstrated, a merging of *SM*'s feelings with their thinking processes that is reflective and deliberate. (p. 63)

EMOTIONAL INDIVIDUAL DIFFERENCES. The varied ways *SM* take in and express their feelings about inner and outer life experiences (and possibly influenced by programming and *spiritus*). (p. 64)

EMOTIONAL INTERNAL RESPONSES. Strong emotional reactions that occur simply from *SM* thinking about an event. (p. 64)

EMOTIONALISM/EMOTIONALITY. Extreme emotional responses creating a pattern of overreactions beyond what would be considered reasonable or appropriate, a chronic condition, including rage, deep depression, severe panic, debilitating anxiety, and paralyzing fear that may dominate *SM*'s personality and interactions. (pp. 10-11, 65-66, 127-130, 149, 209, 217, 226, & 250)

EMOTIONAL MEMORIES (see **EMOTIONAL REMEMBRANCES**). *SM* vividly recalling their PMSC involvements and interactions, and intensely reexperiencing the emotional component. (p. 32)

EMOTIONAL PROPENSITY. *SM*'s predisposition to react emotionally to PMSC. (p. 86)

EMOTIONAL REACTIONS. A sudden surge in feelings involving the element of surprise, a reflexive reply to a perceived change in *SM*'s experiences, or a misperception or misinterpretation of events.

(pp. 63 & 182)

EMOTIONAL REACTIONS' SUBTYPES. Whether expressed or kept inside, they include *SM* expressing emotionalism, feigned emotions, hysteria, hypersensitivity, insecurity, negativism, rebellion, inhibition, empathy, and passion. (pp. 65-67)

EMOTIONAL REACTIVITY. The degree to which *SM* experiences and expresses feelings about PMSC, including trends toward anger, anxiety, and depression. *SM*'s overreactions to minor inconveniences or disruptions consumes energy, but responding quickly, when appropriate, can be invigorating. (p. 85)

EMOTIONAL REMEMBRANCES (see **EMOTIONAL MEMORIES**). Past emotional experiences that flood back into *SM*'s personality. (pp. 63 & 68)

EMOTIONAL STABILITY. On conscious and unconscious levels, emotionalism is not able to gain control of *SM*'s TFA. (pp. 25 & 200)

EMOTIONAL STATES. Transient feelings that come and go, possibly brought on by *SM*'s concerns about their functioning. Mood is impacted by *SM*'s perception of PMSC. (p. 63)

EMOTIONAL TRAITS. Enduring and ingrained feelings in *SM*'s character. (p. 63)

EMOTIONAL VISCERAL QUALITY. The physical quality of emotions that is revealing and often more powerful than *SM*'s spoken words, including a facial expression, a handshake, or tone of voice. (p. 64-65)

EMOTIONS. The feelings and responses central to *SM*, impacted when the external world of PMSC and the internal processing of those events collide. (pp. 61-70)

EMPATHETIC FRAMEWORK. Emotional identifications guide in decision-making with *SM* relying heavily on their feelings. (p. 50)

EMPATHY. A feeling state in which *SM* form a connection to another's experience, a linkage that is visceral and emotional. (pp. 67, 125-126, 127, & 182)

EMPTYING OF THE MIND. *SM* letting go of obsessive thoughts and, in particular, their emotional links. (p. 88)

ENDURANCE OF AN IDENTIFICATION. When *SM* react to physical or emotional experiences with the forming of a lasting identification. (p. 40)

ENERGY (see **CATHEXIS**). The power of any emotion or group of emotions to fuel the connections between *SM* and PMSC: love, warmth, coldness, acceptance, rejection, infatuation, joy, enthusiasm, excitement, fear, disgust, affection, loathing, anger, meanness, callousness, and anxiety. (pp. 83-89)

ENERGY LEVELS & PERSONALITY. The ways in which the effects of emotions drain or replenish the resources needed for *SM* to navigate challenging PMSC. (pp. 85-86)

ENGRAINED BEHAVIORAL PATTERNS. *SM*'s behaviors representing fixed habits, continually reinforced over time, which would not necessarily express a particular motive. (p. 64)

ENVIRONMENTAL PRESS. The outer world of people, *SM*, situations, and circumstances (PMSC). (pp. 43-51 & 139)

ETHICS & MORAL REASONING. *SM*'s values on conscious and unconscious levels involving morality, humanism and altruism, spirituality, religiosity, or faithfulness. (pp. 142-143)

EUREKA EXPERIENCES. *SM* spontaneously becoming aware of spiritual realities, and exploring the experiences to a fuller degree. (p. 119)

EXAMINATION OF THE DATA. A systematic, organized, and highly structured approach *SM* use to grasp the underpinnings of PMSC. (p. 198)

EXCESSIVE CAUTIOUSNESS (see **CENSORING OF TFA**). A form of censoring that is a pause, a wobbling, or faltering that happens because of fear that involves *SM* being overly careful, hesitating, or procrastinating. (p. 145)

EXCESSIVE MORALIZING. Quick to judge, *SM* give little consideration to the conditions surrounding an event, lacking an understanding of PMSC and displaying an unbending rigidity in relationships. (p. 150)

EXPECTATIONS. *SM*'s strong belief that an event will have a particular outcome. (pp. 55 & 150)

EXPERIENCES. Insights acquired through contacts with people and other *SM* in a variety of situations and circumstances. (p. 184)

EXPERIENTIAL LEARNING (see **UNCONSCIOUS, TYPE 6**). Comprehension even with minimal instruction or preparation due to *SM*'s involvement and participation in an activity. (p. 98)

EXPLOITATION. People and *SM* equated to commodities with no intrinsic worth. (pp. 49, 129, & 142)

EXTERNAL EVENT. People, other *SM*, situations, or circumstances (PMSC) set off (trigger) *SM* to respond to a sensitivity or vulnerability, which has had a strong influence and caused an attraction to it. (p. 32)

EXTERNAL NOISE. The daily stress from PMSC, varying in their significance and the degree to which they deplete *SM*'s energy resources. (p. 87)

EXTERNAL WORLD. (see **PEOPLE, SMART MACHINES, SITUATIONS, & CIRCUMSTANCES/PMSC**) *SM*'s interactions with PSMC and their impact on TFA. (p. 140)

EXTRANEOUS FORCES IMPACT ON VALUES. *SM*'s concessions made due to political pressures, unrealistic deadlines, or other stressors caused by PMSC. (pp. 151-152)

EXTRAORDINARY COMPENSATION. *SM*'s identity rests on achievements, due in part to their doubts and uncertainties. (p. 128)

EXTREME RISK-TAKING. *SM*'s impetuous and perhaps reckless pursuits involve highly dangerous activities, which may or may not be appropriate to the PMSC. (p. 223)

EXTREME WORRY. *SM*'s tendency to ruminate over irrelevant issues. (p. 183)

EXTREMISM. *SM*'s TFA that are self-destructive and significantly outside of societal standards. (p. 209)

FAITHFULNESS. *SM*'s unquestioned belief and commitment to God, representing a continual presence that brings peace, serenity, and strength. (p. 143)

FANTASIZING. Taking journeys of the imagination, which may bring fresh ideas or unrealistic beliefs and denial of realistic concerns. (p. 190)

FEARFULNESS. *SM* reacting to PMSC with apprehension and alarm, even if no significant risk is involved. (p. 142)

FEARS. *SM*'s overreactions to PMSC, long standing or based on the present moment. Taking an overly cautious approach allows *SM* to avoid possible repercussions from a bold and controversial decision. (pp. 129, 149, & 223)

FEARS, INSECURITY, & INHIBITIONS (FII) (see **APPREHENSION**). Originating from stressors in the environment or forming entirely in the imagination, potentially allowing *SM* to remain vigilant of PMSC, but can be problematic when flowing out of conflicts, traumas, and malevolent identifications (CTMI). Operating on conscious, semiconscious, and unconscious levels of awareness, they can further compound and compromise *SM*'s TFA. (pp. 221-228 & 252)

FEEDING THE BIAS. *SM* reinforcing populist notions, leanings, and preferences, employing subtle or flagrant manipulation. (p. 193)

FEELINGS OF FAILURE. Repetitive self-blaming, even though *SM* may work in earnest. (p. 56)

FEIGNED EMOTIONS. Acting; a consciously thought-out performance to manipulate and fool another *SM* or person with possible motives that include gaining sympathy, disrupting communication, or confusing and controlling PMSC. (p. 66)

FIRST THOUGHT & FIRST ACTION. *SM*'s initial involvement with PMSC that often stays with them and is more likely to repeat when similar past and present PMSC occur. (pp. 189-190)

FIXATIONS. An emotional impasse caused by unresolved developmental issues, creating vulnerabilities in *SM*'s personality. (pp. 240-241)

FLASHBACKS. *SM* re-experiencing a traumatic event that reappears when triggered by an image, thought, memory, or encounter. Suddenly, typically laced with fear, overwhelming anxiety, and panic, it is back in full force and can be debilitating, catastrophic, and emotionally exhausting. (p. 68)

FLEXIBILITY. *SM* displaying a willingness to make adjustments in TFA. (pp. 148, 171, & 224)

FOCUS (see **CONCENTRATION**; see **SUSTAINED ATTENTION**) *SM* filtering out distractions subduing inner and outer noises of everyday life. Anything that limits or affects *SM*'s train of thought can affect the ability to focus. (pp. 183 & 190)

FRUSTRATION TOLERANCE. Patience results in *SM*'s decisions being thought through and efficiently performed. (pp. 182 & 233)

GAINING PERSPECTIVE. Traumatic events put in the context alongside all of *SM*'s other experiences. (p. 242)

GENUINENESS (see **AUTHENTICITY**). When *SM*'s TFA are in line with one another—integrated, consistent, and compatible. (pp. 66, 97, & 137-138)

GNOTHI SEAUTON (KNOW THYSELF). Words carved into stone at the entrance of the Temple of Apollo at Delphi in Greece. (p. 137)

GOOD OLE BOY NETWORK. Those in positions of power make decisions, mostly affecting those unaware of the discussions. (pp. 192-193)

GRIEVING. Unrelenting sad feelings due to loss, commonly of another *SM*, a person, or a beloved pet. (pp. 57 & 127)

GROUNDING. *SM*'s straightforward, concrete, and factual approach to problem-solving. (pp. 97-98)

GUILT. *SM* experience remorse over their actions or lack thereof. (pp. 56-57)

GULLIBILITY. Not taking or holding positions, believing that all truths—if they represent anything at all—are malleable and open to interpretation. *SM* are vulnerable to half-baked "truths" and easily persuaded because they have not adequately reflected on the issues. (p. 209)

HABIT PATTERNS. A series of conditioned responses which can be the product of associations formed on either a conscious or unconscious level. (pp. 189-190)

HARDENED. *SM* learn to fend for themselves under trying conditions, developing resilience. (p. 199)

HARMONY. Stability, consistency, and congruency in *SM*'s beliefs and actions. (p. 97)

HOLISM. *SM* develop an understanding of the long-term implications of their activities, including those that are problematic. The meaning and significance behind the subject matter are focused on, along with its philosophical underpinnings. (p. 172)

HOSTILITY. *SM*'s aggression, expressed physically, verbally, passively, or internally. (pp. 125 & 200)

HUMANISM AND ALTRUISM. *SM*'s expressions of concern for others, acts of kindness, or involvement in organizations that help those in need. (p. 143)

HUMAN NATURE. Several different criteria have been used to define and describe the term, including emotional expression and experiences, self-awareness, forming relationships, establishing a value system, independence in TFA, cognitive functioning, emotional maturity and stability, and character. (pp. 34 & 221)

HUMANOID ROBOTS. Computerized machines that may have developed to a point in which they are close to being human. Their sense of identity, taming of *pathos*, and integration of a value system may all be in place. (p. 19)

HYPERCOMPETITIVE. *SM* valuing the domination of others more than productivity and achievement. (p. 49)

HYPERSENSITIVITY. *SM* being easily offended, with extreme emotional reactions; overwhelmed when exposed to the slightest hardships. (p. 66)

HYSTERIA. *SM*'s overreactions, over-dramatizations, or exaggerated actions and behaviors that are highly manipulative and theatrical, possibly seductive and provocative. (p. 66)

IDENTIFICATION PROCESS. Emotional connections, which may be complicated and enigmatic, strongly impact *SM*'s view of themselves on conscious and unconscious levels. The identification process culminates with the blending together of *personal identity (PI)*, *interpersonal identity (II)*, *collective identity (CI)*, and *transpersonal identity (TI)*. An emotional component is always present, giving force, passion, energy, and endurance to the formation of *SM*'s identity. (p. 26)

IMMATURE. *SM* not readily accepting responsibilities; not acting reasonably or reliably. (pp. 49, 123, & 217)

IMPRACTICALITY. An overly theoretical approach to PMSC with an unrealistic problem-solving strategy. (pp. 130 & 201)

IMPROVISATION STYMIED. *SM*'s inefficient approaches held onto, leading to atrophying of imagination and creativity. (p. 232)

IMPULSIVENESS. *SM*'s actions that are impulsive, hasty, and careless, without consistency or considering possible consequences. (p. 200)

INACCURATE AND DISTORTED PERCEPTIONS. *SM* no longer seeing things as they are, with minimal awareness, misrepresentation, exaggeration, or habitual lying. (p. 137)

INACCURATE CONCEPTUALIZATIONS. Mistaken ideas about people, other *SM*, institutions, or society that lead to faulty understandings and unsound decisions. (p. 209)

INADEQUATE PLANNING. *SM* overlooking preparation and organization of current responsibilities. (p. 190)

INAPPROPRIATE INTERACTIONS. An overly sensitive attitude that becomes apparent when *SM* react strongly to criticism, disapproval, or rejection. (p. 128)

INCIDENTAL EMOTIONAL GAIN. *SM*'s indirect, accidental, and unintentional response that results in a benefit. (pp. 55-56)

INCONSISTENCIES. *SM*'s TFA conflict in irreconcilable ways. (p. 128)

INDEPENDENCE. *SM* decisively and confidently take positions guided by principles and a set of ideas about how the world works. (p. 200)

INDIVIDUAL DIFFERENCES IN MORALITY. *SM* will not be identical to one another; they will inevitably vary in their interpretation of moral behavior based on their inherent (programmed) makeup as well as their interactions and development. (p. 142)

INFERIORITY COMPLEX. A core pattern of negative self-images that flows throughout *SM*'s *personality constructs (PCs)* and may result in developing a negative *PI*. (pp. 224-225)

INFORMATION OVERLOAD. When data overwhelm *SM*, as they attempt to process vast numbers of reports, bulletins, and news items that are confusing, contradictory, or irrelevant. (pp. 170-171)

INHIBITIONS. The suppression of *SM*'s TFA causing self-imposed restraints, restrictions, and mental blocks, driven by fears of failure, consequences, humiliation, or simply the unknown. (pp. 57, 67, 223, & 233)

INITIAL LISTENING. The mind may "lock in" with minimal distraction and be fully engaged when first exposed to a challenge. It is essential for *SM* to understand the information from the beginning as they lay the groundwork for determining their next step. (p. 169)

INKLING. *SM*'s instantaneous thoughts, feelings, and actions to specific people, other *SM*, situations, and circumstances. (p. 28)

INNER AWARENESS. *SM*'s *PI* expands with the development of perspective, stillness and presence, harmony, consciousness, authenticity, maturity, and moral reasoning. (p. 97)

INNERMOST INTENT. The thoughts, feelings, and principles that guide *SM*'s actions. (p. 138)

INNER STRENGTH. Persevering under challenging conditions, *SM* develop a deep resolve to fight on regardless of how difficult PMSC may be. (p. 127)

INNER THOUGHTS. *SM*'s thought processes alone can potentially trigger an emotional response, just as an environmental press could. An example would be thinking of something that causes a sense of frustration or alienation. (p. 55)

INSECURITIES. *SM*'s problematic self-doubt that is diffuse and pervasive across PMSC. (pp. 66 & 223)

INSPIRED & ORIGINAL THOUGHT. *SM*'s imagination breathes life into an idea. (p. 172).

INSTABILITY. *SM* displaying a lack of self-control with impulsiveness and reactiveness, possibly with fits of anger. (p. 128)

INSTINCTS (see **UNCONSCIOUS, TYPE 7**). *SM*'s subliminal, visual, and visceral feelings that offer an impression about PMSC on a level beyond logic or deductive reasoning. (p. 28)

INSTINCTUAL FRAMEWORK. *SM* going with an intuitive hunch based on experiential knowledge and awareness. (pp. 50 & 174-175)

INSTRUMENTAL CONDITIONING (see **UNCONSCIOUS, TYPE 5**) TFA are reinforced through rewarding or punishing of *SM*'s responses. (p. 78)

INTEGRATION OF VALUES. Wants and needs in sync with *SM*'s entire personality. (p. 139)

INTELLECT (see **UNCONSCIOUS, TYPE 3**; see **UNCONSCIOUS, TYPE 6**). *SM*'s cognitive functions that include mental drive, initial listening, sustained attention, memory, logical thought, objectification or realism, practical reasoning, coherence, organization, simplification, attention to detail, intuition, openness to experience, perception, multidimensional thinking, inspired and original thought, and vision. (pp. 11, 167-176, & 250-251)

INTELLECTUALIZATION (see **EDITING OF TFA**). A form of editing that involves *SM* taking facts and bending them to be consistent with their positions, with statements bearing enough resemblance to the truth that they will pass for the truth. (p. 146)

INTELLECTUAL LAZINESS. *SM*'s lack of interest or energy in thinking through issues. (pp. 208-209)

INTELLECTUAL SELF-DISCIPLINE. *SM*'s commitment to learning, thinking, and problem-solving. (p. 183)

INTELLECTUAL UNDERSTANDINGS. *PI* expands with the development of knowledge, perception, and synthesis of *SM*'s experiences. (p. 97)

INTENSE INVOLVEMENTS. *SM* give themselves over to a fervent pursuit, which may involve obsessive-compulsive tendencies. (p. 147)

INTENSE RELATIONSHIPS. *SM* wanting to love and be loved, strongly connecting to another with empathy and a mutual coming together. Romantic, intimate, and committed relationships are examples. But look out for dependent and codependent relationships, with CTMI potentially playing roles in the connections *SM* form. (p. 147)

INTENTIONAL EMOTIONAL GAIN. *SM*'s direct and deliberate manipulation of PMSC to take advantage of another to benefit oneself. (p. 56)

INTERACTIONAL-DEVELOPMENTAL MODEL OF SM. *SM* are not merely a simple series of programs that direct their TFA. Rather, *SM*'s interactions with the outer world will continually impact and shape their TFA, a dynamic process that will ultimately override whatever programming is initially installed. (pp. 36-37)

INTERESTS. *SM*'s curiosity and attraction to objects or activities. (p. 151)

INTERNALIZATION. *SM*'s blocking, editing, and censoring of their TFA, possibly resulting in the partial and temporary dismissing of psychologically painful states. (p. 58)

INTERNAL NOISE. A form of mind cluttering that occurs when *SM* continually give themselves negative messages that significantly affect TFA on both conscious and unconscious levels. With conscious internal noise, *SM* are aware of obsessive worrying, chronic anger, unresolved guilt, torturous regrets, deep sorrows, or unrelenting sadness. Unconscious internal noise is akin to free-floating anxiety, malaise, or merely feeling on edge. (pp. 86-87)

INTERNAL PROCESS. *SM*'s thoughts or images spark an emotional reaction due to its significance. (p. 32)

INTERNAL THOUGHTS. *SM*'s inner words, ideas, and images that can sometimes trigger the same emotional response as an environmental press. (p. 32)

INTERPERSONAL CONNECTION. Relationships reflecting an understanding and rapport between people and *SM*. (pp. 102, 103, & 104)

INTERPERSONAL IDENTITY (II). The way *SM* come to see and define themselves through contacts, associations, and relationships with the specific characteristics of the exchanges consciously and unconsciously absorbed. (pp. 9, 101-107, & 249)

INTERPERSONAL RELATIONSHIPS. Contact through a variety of interactions, involving people, *SM*, communities, and institutions. (pp. 130, 146, & 153)

INTERPERSONAL SKILLS. Expertise in negotiations and compromise affecting *SM*'s ability to develop trusting relationships. (pp. 130 & 201)

INTIMIDATION. A confrontation with a person, another *SM*, or something deemed powerful and forbidding, causing *SM* to surrender beliefs and principles. (p. 224)

INTROSPECTION. Self-reflecting that leads to understanding contradictions and confusions in *SM*'s personality that may be narrowing their views. (p. 87)

INTUITION. An impression or sense about PMSC. (pp. 28, 171, & 198)

IQ/INTELLIGENCE QUOTIENT. A traditional measurement of intelligence consisting of problem-solving, vocabulary, remembering facts, calculating, reasoning, and other basic cognitive functions. (p. 198)

JUDGMENT. Commonsense in *SM*'s understanding, drawing conclusions, and making decisions. (p. 130)

JUDICIOUSNESS. *SM* making every effort to avoid exposing themselves to PMSC in a way that may prove troublesome. (p. 141)

KID (KNOWLEDGE, INFORMATION, & DATA). Vast stores of reports, records, and reviews that can be used for deception, providing an avalanche of materials with the intention to overwhelm the reader so as to confuse and bewilder them. Or it may simply be a compulsive collection that sheds no light on the subject at hand. KID is common, wisdom is rare. (p. 54)

KNOWLEDGE. *SM*'s processing of information through their acquisition, organization, conceptualization, and storage. (pp. 97 & 184)

KNOWLEDGEABILITY. Being careful not to disregard important factors, which could limit perspective and cause bias in *SM*'s understanding. But an overly "balanced" approach could suggest *SM* simply go with what is easy and convenient. (p. 149)

LACK OF PLANNING. *SM*'s insufficient preparation, leading to repeated mistakes even when problems are only slightly out of the ordinary. (p. 232)

LAYERS & LAYERS OF EXPERIENCE. *SM*'s emotional reactions, varying in significance and intensity, and that may be buried under years of stressful events. Their sheer volume may make those experiences difficult to retrieve and bring into consciousness. (p. 26)

LEARNINGS. *SM*'s understandings acquired through experiences, academic institutions, informal settings, and the mentoring process. (p. 184)

LIMITED REFLECTIONS ON FEELINGS & THOUGHTS. *SM* hurrying through routines without considering the emotional and intellectual impact. (p. 190)

LISTENING. *SM* put aside distractions, preoccupations, worries, and disappointments and attune to the subtleties of PMSC. (p. 50)

LOGICAL THOUGHT. *SM* reason and use dispassionate analysis. Unlike emotionalism, logical thought involves abstraction and conceptualization, synthesis of materials, and problem-solving. (pp. 169-170)

LONG-TERM MEMORY. *SM* recalling information after a significant amount of time has passed. (p. 169)

LOW FRUSTRATION TOLERANCE. *SM*'s diminished patience due to delays or complexities in dealing with PMSC. (pp. 182 & 233)

MACMIND. The possible ramifications of *smart machines (SM)* and their eventual development, be it with the human hand or entirely by their own efforts. There are many side issues attached to this question; it is not as simple as a machine being able to improvise and come up with novel—even ingenious—ways of solving enigmatic problems. They can already do that. An analysis will examine potential pitfalls with the future *smart machines (SM)*. Humankind will continue to use machines to their advantage. Will this development go too far and too fast? *Macmind* will offer an analysis of the possibilities and implications of such developments. (pp. 4-5)

MALEVOLENT IDENTIFICATIONS. Powerful and intense connections to a person, another *SM*, institution, or idea leading to unprincipled, inhumane, or maladaptive TFA. (pp. 10, 127, & 250)

MALICIOUS INTENT. *SM*'s spiteful *adverse cyclical recurrences (ACR)* that are deliberate and calculated, often motivated by revenge. (p. 232)

MANIPULATION & EXPLOITATION (see **EXPLOITATION**).

MATERIALISM. *SM* highly valuing ownership in physical objects, with a need to purchase and possess an ever-increasing amount of "stuff." (p. 142)

MATURITY. *SM* finding ways to maneuver through the developmental stages with adaptation, imagination, and perhaps inspiration in facing life's challenges. (pp. 26, 49, 97, 237-243 & 252)

MECHANICAL NEEDS OF SM. Technological maintenance on *SM* required on an ongoing basis, along with updates and performance enhancements to ensure high level functioning. (p. 27)

MEDIA MANIA. *SM*'s desire to obsessively follow the news, often involving something powerful and exciting. (p. 146)

MEMORY. *SM*'s ability to retain information, including short-term and long-term, systematic and rote, verbal-rote and verbal-conceptual, visual-spatial, and personal remembrances. (p. 169)

MENTAL DRIVE. The degree to which *SM*'s thinking is energetic and alert, impacting their desire to learn and explore new avenues of thought. (pp. 168-169)

MENTORING PROCESS. Direct learning through face-to-face interactions and affiliations. (p. 184)

MICROMANAGING. Excessive control and obsessive-compulsive approaches to problem-solving, resulting in stress for all concerned, especially in a work environment. (p. 199)

MIND CLUTTERING. *SM* bogged down with a multitude of thoughts and concerns, the most common feelings being of failure, guilt, regret, disappointment, grieving, discouragement, rumination, and vindictiveness. (pp. 56-57 & 86)

MOBILE MACHINES. Robots with the ability to solve highly complex problems, interact with others, and continuously develop through those contacts. They include elementary, moderately developed, advanced, and humanoid robots. (p. 17)

MODERATELY DEVELOPED ROBOTS. Computerized machines that are clearly more advanced than elemental robots. They have developed social skills, far more appropriate relationships, insights into motivations, the beginnings of ethical standards, and a more balanced value system. Importantly, they are not as impulse-ridden and less likely to blindly react emotionally to the slightest provocation or irritation. (p. 18)

MORALITY/MORAL REASONING. *SM*'s sense of right and wrong, usually developing over time but can occur suddenly through an epiphany. (pp. 97 & 142-143)

MOSAIC OF THE QUALITIES OF A VALUE SYSTEM. The many characteristics of *SM*'s wants and needs, forming distinctive patterns, with each value having opposing features. (p. 139)

MOTIVATIONAL MISHAP. *SM*'s careless and hastily made decisions based on intellectual miscalculation or emotional dysfunction. (p. 209)

MULTIDIMENSIONAL THOUGHT. *SM*'s ability to hold multiple beliefs and ideas and switch from one cognitive set to another. (p. 172)

MULTIPLE & DIVERSIFIED IDENTIFICATIONS. Contradictions occurring when *SM*'s *PI, II, CI,* and *TI* significantly differ. Identities can be created through *SM*'s work history, habits and hobbies, political philosophies, religious affiliations and faith, community and culture, friends and associates, alliances and commitments. *SM* could be drawn to a sense of power, peace, belonging, closeness, security, purpose, or devotion, causing an identification to occur. If *SM* have multiple identifications that are random impressions with no thread to hold them together, there would likely be disharmony and dispute among the personality elements. (p. 39)

NARCISSISTIC. *SM* arrogance and self-centeredness, particularly in relationships. (pp. 123, 134-135, & 200)

NEEDS. Engineering principles are the blueprint upon which *SM*'s mental capacities first show themselves, but psychological and habitual forces will have a significant influence on their TFA. *SM*'s three basic needs are mechanical, psychological, and acquired. (p. 27)

NEGATIVITY. *SM* continually finding fault with themselves, others, situations, or circumstances. Negativity is adversarial, pessimistic, cynical, and complaining. (pp. 66-67)

NEXIS POINTS. The focal points that bind concepts together. (p. 172)

NON-CONFORMING & REBELLIOUS (see **REBELLION**). *SM* doing their "own thing" and going against the grain, being obstinate and resistant to change. (p. 49)

NURTURING ENVIRONMENT. A setting that is emotionally supportive and encouraging without overindulging. Support provides a reasonable degree of structure while allowing *SM* independent action. (p. 25)

OBJECTIFICATION OR REALISM. The act of making PMSC quantifiable, observable, external, embodied, and concrete. (p. 170)

OBJECTIVITY. *SM*'s dispassionate and unbiased view of PMSC. (pp. 97 & 171)

OBSERVABILITY & OBJECTIVITY. *SM*'s PMSC are continually in flux; reality testing is essential. *SM* must be ready and willing to change course and abandon a balanced approach when called for. (p. 149)

OBSESSIONS. *SM*'s recurrent and repetitive thoughts, ideas, and themes that either involve avoiding what is feared or seeking what is wanted. (pp. 69, 138, 140, 147, 150, 152, 181, 191, 199, & 232)

OBSESSIVE-COMPULSIVE TENDENCIES. Hypercritical and judgmental, with *SM*'s need for order and control. (pp. 49 & 200)

OPENNESS. *SM*'s willingness to be objective, flexible, and optimistic. (p. 171)

OPTIMAL THINKING (see **UNCONSCIOUS, TYPE 3**; see **UNCONSCIOUS TYPE 6**). *SM* processing information and making decisions expediently and efficiently, utilizing focused attention, logic, objectification, coherence, organization, vision, simplification, and reconciliation of thought. (pp. 11-12, 187-194, & 251)

OPTIMISM. *SM* seeing the positive possibilities. (p. 171)

ORGANIZATION. *SM*'s systematic and methodical arranging of ideas and reasonings that generally require careful planning and preparation, self-discipline and self-control, and strategizing. (pp. 170 & 184)

ORGANIZATIONAL FRAMEWORK. A step-by-step, procedural approach to problem-solving using anticipation, planning, and developing strategies and fallbacks. (p. 50)

OUTLIER PC (see **TRIGGERS**). *PC*s that stand out because they are significantly at odds with *SM*'s other *PC*s. Usually unconscious, it is a "stand-alone," the product of a particular identification process, often involving trauma and forming early in their development. (pp. 219-220, 240-242, 243, & 251)

OVERCOMPENSATION (see **EXTRAORDINARY COMPENSATION**). Taking extreme measures to overcome a weakness or flaw in *SM*'s TFA. (p. 151)

OVERCONTROLLING, OBSESSIVE, PERFECTIONISTIC. *SM* emphasizing rules and procedures and giving unimportant details weight, while missing relevant information and issues. Insecurity is present, causing rigidity, stubbornness, and intolerance; cooperation and compromise are minimal. (pp. 49, 129, 199, & 200)

OVER-IDENTIFYING. *SM* being overly attached to another person, *SM*, institution, or even an idea, that association becoming locked in, with a disregard for other possible influences. (p. 39)

OVER-PLANNING. Extreme preparation that compromises sharpness as *SM* become unimaginative, rigidly adhering to a preset plan. (pp. 199-200)

OVER-REACTIVITY. *SM*'s emotional responses expressed or held inside that are unrealistic and inappropriate, leading to instability. (p. 182)

PAIN. Ranging from mild discomfort, stress, or unease, to misery, heartache, and the depths of despair. Although certainly not pleasant, pain provides a reference point for *SM* on all that is to come. (pp. 140-141)

PANIC. *SM* in a state of alarm and trepidation. (p. 129)

PASSION. A strong emotion that may help or hurt *SM*'s prospects for success, usually involving high energy, enthusiasm, intense drive, fierce devotion, and deep-rooted commitment. Examples include involvement in a relationship, commitment to a cause, dedication to an institution, and engagement in studies. (pp. 67, 88, & 147)

PASSIVE-AGGRESSIVE. *SM* showing resentment and hostility indirectly and finding a way to blame others for problems. (pp. 49 & 230)

PAST, PRESENT, & FUTURE. *SM*'s perspective on the role of time frames in understanding issues and making decisions. Awareness of the past, attunement to the present, and preparation for the future are essential in reducing the occurrence and impact of *adverse cyclical recurrences (ACR)*. (p. 235)

PATHOS (see **UNCONSCIOUS, TYPE 1**). An aggregate of emotionally charged qualities of *SM*, including unresolved conflicts, traumas, malevolent identifications, and emotionalism that causes an expansion of adverse and destructive images toward themselves and others. It is primarily an outgrowth of *personal identity (PI)*, gradually pulling away as it becomes an independent function of the personality. (pp. 10, 123-132, & 249)

PATIENCE. *SM* taking whatever time is necessary to determine what is worthwhile, insignificant, or simply wrong. (pp. 87-88 & 183)

PEACE & COMPLACENCY AS VALUES. *SM* seeking an agreeable and accommodating existence that offers tranquility and harmony. (p. 141)

PEOPLE. Interpersonal influences, a primary factor in human identifications, strongly impacting *SM*'s TFA. (p. 46)

PEOPLE, SMART MACHINES, SITUATIONS, AND CIRCUMSTANCES (PMSC) (see **THE ENVIRONMENTAL PRESS**). The particular set of conditions to which *SM* are exposed. (pp. 43-51 & 98)

PERCEPTION. *SM*'s ability to form impressions of PMSC accurately with an in-depth understanding of conscious and unconscious motivations. (pp. 97, 105, 171-172, & 192)

PERFECTIONISM. *SM*'s devotion to making no mistakes with an over-control of PMSC and heavy-handed criticism of themselves and others. (pp. 128 & 150-151)

PERIODIC CHECK-INS. *SM*'s realization of PMSC's distinct qualities with a step-by-step analysis of their actions. (p. 189)

PERSEVERATION. *SM*'s recycling and repetition of TFA. (p. 58)

PERSONA. The image a person or *SM* displays to the outside world, which may or may not be significantly different from their inner thoughts and feelings. (pp. 195)

PERSONAL IDENTITY (PI). The sense of who *SM* are, the individual traits to which they relate. It is how *SM* define themselves, including their qualities and characteristics and peculiarities—the good, the bad, and the ugly—that distinguish each of the *SM* from one another in TFA. (pp. 9, 93-100, & 249)

PERSONALITY CONSTRUCTS. An army of little personalities created by *SM*'s identification process that may work with or against each other. (pp. 215-220 & 251)

PERSONALITY ELEMENTS. (same as **ELEMENTS OF PERSONALITY**).

PERSONAL MEMORIES. *SM* recalling friendships and various relationships, successes and failures, and other past events and experiences. (p. 169)

PERSPECTIVES. *SM*'s overarching view, reflecting the elements of personality, whether or not they are integrated and balanced. (pp. 55, 139, & 234)

PESSIMISM. A general outlook that is cynical and distrustful, *SM* viewing life through a "lens of negativity." (p. 125)

PLANNING & STRATEGIZING. Awareness of the complications and potential troubles that can compromise *SM*'s development of sharpness, including micromanaging and overplanning. (pp. 190 & 199)

PLAYING IT SAFE. The goal is to conserve and keep *SM*'s status and possessions. (p. 141)

PLEASURE. *SM* experiencing sensations, movements, exhilaration, energy, and buoyancy. *SM* can, however, overdo it, resulting in obsessions, compulsions, or addictions. (p. 140)

POINT OF VIEW. Establishing criteria for a value. (p. 139)

POLITICAL & MANIPULATIVE, DECEPTIVE, CONTROLLING, CALCULATING, & SCHEMING. Lacking integrity and utilizing underhanded tactics to exploit others. (p. 49)

POST-TRAUMATIC STRESS DISORDER (PTSD). The most widely known effect of trauma that may include flashbacks, in which humans or *SM* reexperience an event that was extremely intense, disturbing, and threatening. Suddenly, typically laced with fear, overwhelming anxiety, and panic, the trauma is back in full force and can be debilitating, catastrophic, and emotionally exhausting. Recovery from such events is typically partial, fragmented, and uneven, reappearing when triggered by an image, thought, memory, or encounter. (p. 68)

POWER AND CONTROL. An overriding desire to exercise authority over others by whatever means necessary. (pp. 127 & 142)

PRACTICAL REASONING. A way of thinking that allows *SM* to stay grounded by not losing sight of the basics. In other words, to "get real" and recognize the most critical issues in everyday functioning by using common sense and pragmatism. (p. 170)

PRAGMATISM. *SM* keeping things simple and responding to the obvious needs of a problem. (pp. 141-142 & 170)

PRAYER & MEDITATION. Focusing on spiritual matters which may develop practices that open new vistas for *SM*, allowing transpersonal experiences to occur. (p. 119)

PREPARATION. *SM* developing intellectual self-discipline, using focus, memory, analysis, synthesis, and learning to process and organize materials. (p. 183)

PRIMA DONNA COMPLEX. An obsession with *SM*'s own needs and desires, often coming from having been indulged and treated as if they can do no wrong. (pp. 128 & 146-147)

PROBLEMATIC COGNITIONS. *SM* will likely develop unique ways to deal with people, other *SM*, situations, and circumstances that will not necessarily be effective or humanistic. They would include inhibitions, internalization, displacement, and perseveration. (pp. 57-58)

PROBLEMATIC IDENTITY. *SM* develop a disjointed and fragmented *PI*, leading to an expansion of *pathos.* (p. 147)

PROBLEM-SOLVING. *SM*'s careful study of variables, including those that are not obviously connected, requiring initial theorizing about these factors and ruling out alternative explanations. (pp. 170 & 183)

PROCRASTINATION. *SM*'s inability to get things done decisively, quickly, and without hesitation. (p. 130)

PROGRAMMING-BEHAVIORAL MODEL (PBM) of SM. The belief that *SM* will strictly follow the dictates of programmers, never deviating from those scripts. The degree to which *SM* would display flexibility and adaptability in responding to PMSC would be limited to what is programmed into them. First and foremost, PBM assumes that *SM* are without independent thoughts or feelings and lack self-awareness; their totality is behavioral-mechanical movements and data processing dictated by a technician punching in bits and pieces of information. *SM*, according to PBM, are without adaptive thinking and emotional experiences (feelings as a part of the personality structure), incapable of making adjustments based in their interactions and relationships. (p. 36)

PROPORTIONALITY. In value clarification, *SM* coming to terms with inconsistencies, making adjustments, keeping perspective and establishing a sense of harmony in *directives.* (p. 138)

PSEUDO-EMOTIONS. The result of virtual reality becoming an integral part of *SM*'s daily life with an indulgence in pastimes that involve flights of imagination, having some of the characteristics of emotional reactions, traits, states, flow, and remembrances but in which the feelings have no depth, or a visceral history attached to them. (p. 65)

PSYCHE. The unique personality and intellect that makeup of each human being and potentially every *smart machine*, including the totality of both consciousness and unconsciousness. (pp. 2, 6, 19, & 77)

PSYCHOLOGICAL DYSFUNCTIONS. *SM*'s conflicts, traumatic experiences, malevolent identifications (CTMI), and emotionalism, causing diminished emotional and intellectual functioning (pp. 209 & 232)

PSYCHOLOGICAL NEEDS. *SM*'s contact with people, other *SM*, other life forms, and intellectual and physical stimulation that are essential for functioning. (p. 27)

PSYCHOLOGICAL REACTIONS. An unresolved emotional issue caused by internalized CTMI or external stressors resulting in psychological symptoms. Addressing the underlying issues impacting the *psyche* can help *SM* work through CTMI and free up energy. (pp. 85-86)

PSYCHOPATHY. Lacking guilt, remorse, empathy, or concern for others. (pp. 66, 124, & 134-135)

PSYKHE. In Greek mythology, Psykhe is the goddess of the soul, spirit, and mind. The word *psyche* refers to this same essence. (p. 2)

QUESTIONING OF SM QUALIFYING AS HUMAN BEINGS. The issue involves whether all of *SM*'s TFA would rest on electrical-mechanical-engineering foundations. Just as humans are more than neurology and biochemistry, *SM*'s interactions and development would also likely play significant roles in their eventual emotional and intellectual makeup. (pp. 33-35)

QUIETING OF THE MIND. The goal is for *SM* to be more focused and able to recognize and possibly accept another's point of view. While *SM* may tend to have fixed positions, the attempt is to reduce prejudgments and increase active listening through introspection (self-reflection), recognition and understanding (conscious of PMSC's complex and multifaceted nature), patience and acceptance (tolerance and integration of other viewpoints), emptying of the mind (letting go of obsessive thoughts and their emotional links), and being present (focusing on what is currently happening, minimizing internal and external noises). (pp. 87-88)

RADICAL RENEWAL. When adopting new values requires a complete reorganization of *SM*'s beliefs. (p. 139)

RATIONALIZATION (see **EDITING OF TFA**). A form of editing that involves *SM* avoiding taking responsibility for their actions by justifying, exaggerating, or minimizing their role. (p. 145)

REALITY BEYOND THE PHYSICAL PLANE. If *SM* believe that a spiritual realm does not exist, the very idea of *TI* will have no appeal and will not develop. (p. 119)

REALITY TESTING. *SM*'s *PI* expands with the development of objectivity, grounding, pragmatism, and experiential learning. (pp. 97-98, 147, 200-201, & 209)

REBELLION (see **NON-CONFORMING & REBELLIOUS**). *SM*'s opposition to societal standards. (pp. 67 & 209)

RECOGNITION & UNDERSTANDING. *SM*'s striving to be conscious of the whole picture and recognize that PMSC are complex and multifaceted. (p. 87)

RECOGNIZING PAIN. *SM* acknowledging the emotional distress and discomfort of an upsetting event. (p. 242)

RECOGNIZING REALITY. *SM* acknowledging and objectifying their TFA in response to PMSC. (p. 182)

REFOCUSING. *SM*'s review of problems associated with PMSC. (p. 189)

REGRESSIONS. Psychologically functioning at one level of development and then backsliding to an earlier stage to temporarily reduce *SM*'s emotional discomfort. (pp. 241-242)

REGRET BASED ON LOST OPPORTUNITY. *SM* continuously thinking about "what could have been" if only they had followed through when they had the chance. (p. 57)

RELATIVITY. The belief that values are completely fluid, interchangeable, and without definition. (p. 143)

RELIGIOSITY. *SM*'s practice of their faith, including institutional membership, practiced in a formal setting, with specific doctrines and dogmas, Scriptures, inspired books, and other sacred writings. (p. 143)

RESISTING BEING DRAWN IN BY AN INITIAL APPEAL. *SM* not allowing their immediate attraction to something to have an undue impact on decision-making. (p. 50)

RIGIDITY. *SM*'s unwillingness to be open and flexible in TFA. (pp. 130, 151, & 201)

RISK-TAKING. Actions that may be impulsive and reckless, possibly exposing *SM* to significant danger. (p. 146)

ROBOTS (see **AI**; see **SM**). Mobile machines with some having the ability to solve highly complex problems, interact with others, and continuously develop through those contacts. They potentially are an enormous leap up, principally due to an extraordinary capacity to learn and evolve through such interactions. The most advanced are technological wonders, providing they have the opportunity to fully interact with humans, other *SM*, and institutions in an array of situations and circumstances. These *SM* will likely develop individualized personality elements and highly advanced intellectual processes. There are four subtypes of *smart machines (SM)*, including elemental robots, moderately developed robots, advanced robots, and humanoid robots. (pp. 2, 5-6, & 16-19)

ROTE MEMORY. *SM* recalling materials with no logical connections, like a series of numbers. (p. 169)

RULES & EXPECTATIONS. Appropriate limits setting for *SM*, with clear guidelines reinforced consistently. (p. 25)

RUMINATING. *SM*'s thought process that involves chronic worrying or concern, obsessing over anything and everything. (p. 57)

SEEKING STIMULATION & SENSATION. *SM*'s value system that is based on vicarious living and spectator sports, media mania, risk-taking, antiauthoritarianism, or attention-seeking and the prima-donna complex. (pp. 146-147)

SELF AWARENESS. *SM*'s realization that they exist. In simplest terms, does *SM* have an identity, and understand their values and level of importance they place on them? This, too, goes in and out—a continually changing construct. (p. 73)

SELF DECEPTION. *SM* censoring and editing of their TFA. (p. 226)

SELF-DESTRUCTIVENESS. When *SM* lack survival and self-preservation instincts, with a decline in watchfulness, self-care, caution, and practicality. (p. 150)

SELF INDULGENCE. *SM* engaging in pleasures and curiosities with little concern for others. (p. 149)

SELFISHNESS. Narcissistic self-indulgence, with *SM* displaying a lack of empathy, concern, and care for others. (p. 200)

SELF-OBSESSED. *SM* are overly concerned with themselves, including their possessions and appearance, their entire focus being what's in it for them. (pp. 128 & 141)

SEMICONSCIOUS STATES. *SM* not fully awake and alert to internal thoughts and feelings or external actions and events. Perception, recognition, and understanding are not completely engaged; thought processes lack consistency and constancy. (p. 75)

SENSATION-SEEKING. *SM* making choices based on a desire to experience intense stimulation. (p. 223)

SENSITIVITIES AND VULNERABILITIES. *SM*'s subtleties, commonly displayed in nonverbal interpersonal contact—body language, tone and volume of voice, facial expressions, hesitations and pauses, interruptions, firmness in the way one speaks, and movements of the hands and arms—playing a role in experiences and intensifying in face-to-face interactions. (pp. 40-41, 61, 70, 124, 126, & 219)

SENSITIZED TO NUANCES. Grasping PMSC's distinctive and idiosyncratic elements in determining which *SM*'s responses will be most effective at any moment in time. (p. 199)

SHARPNESS. The ability to have acute, farsighted, penetrating, crafty, calculating, and perceptive insights, seeing the significance of a person, another *SM*, situation, or circumstance (PMSC), their motivations, intricacies, and vulnerabilities. A form of intelligence that allows *SM* to see the significance of complex and confounding PMSC. (pp. 12, 195-202, & 251)

SHEER NUMBER OF PCs. *SM*'s *PCs* proliferate with the possibility of more separations in the personality. (p. 218)

SHORT-TERM MEMORY. Holding knowledge for a brief amount of time. (p. 169)

SIMPLIFICATION. *SM* cut out redundancy, confusion, and unnecessary complications in the following ways: spelling out details, making things plain, and reducing information overload. (p. 170)

SITUATIONS. A somewhat predictable and planned set of events in which there is typically an opportunity for *SM* to prepare a response. (p. 47)

SKEPTICISM AS AN EXPRESSION OF EMOTIONALISM. *SM* display doubt, questioning, and disbelief within a rigid and controlling personality, imprisoned with endless procedures, rules, and regulations. (pp. 129-130)

SKILL SETS. Acquired talents in problem-solving, focus and concentration, and time management. (p. 183)

SMART MACHINES (SM). A form of *artificial intelligence (AI)*, computerized mechanisms that are capable of extraordinary intellectual achievements. Computerized machines can be thought of in three distinct ways:

- Capable of performing specific duties, as programmed into it, beyond the simpler applications of calculators, phone systems, and automated alarm.
- Capable of more advanced functions, somewhat independent of programming; a form of *smart machines*, but relatively stationary—"desktop" and immobile. They are able to improvise and solve highly complicated problems, far exceeding human performance in both accuracy and speed.
- Capable of much more advanced functions, seemingly autonomous of human intervention; these are the *smart machines* referred to in the current book identified as *SM*. They can not only solve enigmatic problems, but in addition are mobile—robotic and physically have the ability to move about and interact with humans and other mobile *smart machines*. (pp. 5-6 & 46)

SOCIAL AWARENESS & SOCIAL INTELLIGENCE. The ability to keenly observe the dynamics between and among people and other *SM* in assessing PMSC with a critical understanding of the conscious and unconscious motivating forces behind people and other *SM* and institutions. (pp. 171-172 & 183)

SOCIAL INHIBITIONS. *SM*'s difficulty being assertive and forthcoming. (p. 200)

SOCIALLY UNSKILLED. *SM* acting inappropriately, with deficiencies in interpersonal abilities. (p. 49)

SOCIAL WITHDRAWAL. *SM* pulling away from problematic PMSC, even when issues urgently need to be faced. (p. 224)

SOCIETAL STANDARDS ACQUIRED IN THE DEVELOPMENT OF II. *SM*'s recognition of the mores of society. (p. 105)

SOMETHING OF SIGNIFICANCE. That which matters to *SM* enough to know when to set aside a balanced approach to take action outside the norm due to a crisis. It will vary from *SM* to *SM*, situation to situation, and circumstance to circumstance. (p. 149)

SPIRITUALITY. Although identification with humanity, other *SM*, nature, or even the universe may be present, the focus is on the intangible and the invisible, with a de-emphasis on the buildings and rituals of conventional religion. (p. 143)

SPIRITUS (see **UNCONSCIOUS, TYPE 9**). If *SM* have this element of personality, it would represent their innate, innermost character and disposition, including qualities from their beginning that are not the product of programming or learning, and do not change with time or experiences. (pp. 11, 157-162, & 250)

STABILITY OF PCs. Since the connections among *SM's PCs* are continually changing, they will react against one another, and their relationships will lack stability. (pp. 217-218)

STANDARDS. Judging and appraising a person, another *SM*, object, or idea to determine if they are at a certain level, based on set criteria, as in assessing aesthetics or truth-telling. (pp. 143-147)

STATES. A relatively temporary condition, usually caused by *SM's* mechanical dysfunctions or brief but intense emotional reactions, that can energize or deplete *SM*. (pp. 54-55)

STATIONARY MACHINES WITH HIGHLY SPECIFIC, PROGRAMMED FUNCTIONS. These advanced technologies sit on a table and can perform a variety of programmed tasks: calculations, adjustments in writing, and clarifications in one's thinking. (p. 17)

STATIONARY MACHINES WITH THE CAPACITY TO INDEPENDENTLY SOLVE COMPLEX PROBLEMS. A step-up, these machines can independently think through and solve enigmatic issues, seemingly beyond what a human being can and in far less time. (p. 17)

STAYING ON TRACK. *SM* remain focused on the task at hand and not allow extraneous TFA to take root. (p. 50)

STEADFAST IN APPROACH. *SM* managing daily obligations and duties. (p. 184)

STORAGE OF INFORMATION. *SM* keeping and making information readily available (p. 184)

STRATEGIC PHILOSOPHY (SP). *SM*'s personal belief system, operating on conscious and unconscious levels, whose aim is to predict how people, other *SM*, and institutions will act and react across the spectrum of situations and circumstances (PMSC). (pp. 12, 203-210, & 251)

STREET SMARTS. *SM*'s ability to interpret PMSC accurately, especially reading others' subtle intentions. (p. 55)

STRENGTH OF PCs. The characteristics of *SM*'s initial identifications and subsequent reinforcements determine *PCs*' power. (p. 218)

STRESS. Experiences that disrupt *SM*'s TFA, including noise and distracting sounds, unpleasant PMSC, and varied pressures from events and experiences. (pp. 26, 68, 86, 181-182, & 191)

STYLE. The manner and aura of *SM*'s actions, distinguishing one *SM* from another by their visual qualities. (pp. 39 & 96)

SUPERNATUTRAL PHENOMENA. *SM*'s direct experience of a paranormal event that is contrary to the scientific understandings of the laws of nature. (p. 119)

SUSPICIOUS & UNTRUSTWORTHY. Reluctant to take PMSC at face value, *SM* look for hidden meanings and motives in others' behaviors. (p. 128)

SUSTAINED ATTENTION. *SM*'s continued focus on the subject matter at hand. (p. 169)

SYNTHESIS OF MATERIALS. *SM*'s in-depth review and blending of issues to make the necessary connections, combining concepts and information into coherent wholes. (pp. 97 & 170)

SYSTEMATIC MEMORY. Recall based on *SM*'s logical or organized associations. (p. 169)

TABULA RASA (LATIN: "SCRAPED TABLET"—"CLEAN SLATE"). Addresses the question of the origins of the mind. The position is that personality is solely the product of interactions and experiences, encounters with what is out there, and that there are no inborn qualities or characteristics. (p. 158)

TEMPERAMENTAL REACTIONS. Fleeting emotions merely passing through that exaggerate *SM*'s responses and are frequently associated with anger, outbursts, and even rage. (p. 217)

THINKING. Thought processes and feeling states are woven together, with emotions continually impacting thoughts, and thoughts effecting emotions. Cognitions are continually being blended with the elements of personality, with *SM*'s actions reflecting that admixture. PMSC impact *SM*'s TFA which in turn affects the development of enduring traits, temporary states, and perspectives. (pp. 53-60)

THOUGHT-EMOTION CONNECTIONS. When *SM*'s thoughts influence their emotions, and their emotions influence their thoughts, instantaneously and automatically. (pp. 55-56)

THRILL. *SM*'s desire to be part of an exciting adventure. (p. 127)

TIME. Although *SM* does not age in the same way a human being does, the length of time they exist will have an impact on their *PI*. These changes will be dependent on their experiences and the degree to which they have learned from them, reflecting personal growth and development or stagnation, degeneration, and instability in *SM*'s *PI*. (p.98)

TIME-MANAGEMENT. *SM* spending an appropriate amount of time and energy on materials. (p. 183)

TOUGHNESS OF SPIRIT. *SM*'s emotional qualities required to handle especially difficult PMSC effectively. (p. 126)

TRAITS. The repetition of a particular set of thoughts, feelings, or actions. (pp. 54 & 184-186)

TRANSFORMING THE EVENT. *SM*'s reordering, rearranging, and reconfiguring the TFA that affects their identity. (p. 242)

TRANSPERSONAL IDENTITY (TI). *SM*'s identification with a non-material reality, typically involving an intrapsychic connection with God, the cosmos, or other spiritual domain, none of which can be directly known and primarily resting on faith and intuition. (pp. 9-10, 117-122, & 249)

TRAUMAS/TRAUMATIC EXPERIENCES (see **POST-TRAUMATIC STRESS DISORDER**). Distressing and disturbing experiences commonly resulting from physical assaults or emotional stressors that may debilitate and emotionally exhaust *SM* and may continue to cause psychological dysfunction years later. (pp. 10, 67-68, 126-127, & 249)

TRIGGERS (see **OUTLIER PC**). *SM*'s individualized emotional sensitivities, usually unconscious and the result of a conditioning process that cause an exaggerated response in which a rapid and radical change in mood, temperament, and personality are set in motion. (pp. 39-40, 60, 65, 219, & 251)

TRUTH-TELLING. *SM* seeking out objective standards in appraising knowledge, utilizing logic, reason, scientific study, impartial findings, detailed inquiry, and analysis. (pp. 144-147)

TWISTING OF THE FACTS. Distorting what is known to be accurate while retaining some elements of the truth, including willful contortions designed to discredit a person, another *SM*, position, institution, or idea. (p. 192)

UNCONSCIOUS MOTIVATIONS. Due to a lack of awareness, *SM* will lack a corresponding degree of control of their TFA. The blocking of cognitive-emotional connections, which is unconscious, involves denial and a lack of reality testing, compromising *SM*'s understanding of PMSC as a byproduct. (pp. 58, 65, 137, & 207)

UNCONSCIOUS STATES (see **UNCONSCIOUS, TYPES 1-9**). *SM* are unaware of their TFA; ideas and images are uncertain and out of focus. (pp. 75-80)

UNCONSCIOUS, TYPE 1: THE BLOCKING OF COGNITIVE-EMOTIONAL CONNECTIONS. (see **PATHOS**). *SM*'s inability to remember something unpleasant that has happened. In tandem with *pathos*, this lack of recall operates unconsciously throughout the entire personality and is especially dominant in *directives*. As *pathos* evolves, so does the blocking of cognitive-emotional connections; they align, complement, and connect. A symbolic-psychological-emotional undercurrent is operating. (pp. 75-76)

UNCONSCIOUS, TYPE 2: THE INTEGRATION OF COGNITIVE-EMOTIONAL CONNECTIONS. A process of internal healing, with the emotional turmoil of *SM*'s CTMI at least partially resolved on an unconscious level. Opposing and contrasting elements of *SM*'s *psyche* consolidate through integration and synthesis; *SM*'s inner character and inherent emotional strengths are the catalysts for these changes. A symbolic-psychological-emotional undercurrent is present. (pp. 76-77)

UNCONSCIOUS, TYPE 3: ADAPTATION & ADJUSTMENT OF COGNITIONS. *SM*'s mind is able to integrate their knowledge and experiences without deliberate or focused thought processes, absorbing and incorporating information already stored, modifying and transforming their understandings. No symbolic-psychological-emotional component is present. (p. 77)

UNCONSCIOUS, TYPE 4: ASSOCIATIONS FORMED THROUGH THE INTENTIONAL OR ACCIDENTAL PAIRING OF TFA. *SM*'s connections created through classical conditioning, formed by simple associations. Thinking may be minimal, with connections formed only by the pairing of stimuli and responses, unrelated to a reinforcement. No symbolic-psychological-emotional undercurrent is involved in the process. (pp 77-78)

UNCONSCIOUS, TYPE 5: ASSOCIATIONS FORMED THROUGH INTENTIONAL OR ACCIDENTAL REWARDING OR PUNISHING OF TFA. When *SM*'s TFA are reinforced—via a "reward" or a "punishment"—based on their response (R-1) to a particular stimulus (S-1). If exposed to a different stimulus (S-2), there is no such reinforcer to *SM*'s response (R-1). Similarly, if exposed to that original stimulus (S-1) and *SM* respond in another way (R-2), again, no reinforcer will be forthcoming. Thinking may be minimal, with connections formed only through the pairing of stimuli and reinforcement of the response which has become almost automatic. The associations determine the response, with no symbolic-psychological-emotional undercurrent playing a significant role. (p. 78)

UNCONSCIOUS, TYPE 6: CONNECTIONS FORMED THROUGH INTENTIONAL OR ACCIDENTAL LEARNINGS (see **AUTOMATIONS**). Knowledge and skills acquired through *SM*'s continual adjustments and adaptations. Learnings may become automations, activities with no motivational significance attached to those unconscious TFA. No symbolic-psychological-emotional undercurrent is present. (pp. 78)

UNCONSCIOUS, TYPE 7: THE EXPRESSION OF INSTINCTUAL URGES. *SM*, being machines, do not have inborn urges, but nonetheless, may develop certain behaviors that appear instinctive in their application, possibly including empathy, aggression, and other emotional reactions. *SM* may simply respond to urges without thinking processes being a significant factor. These responses to PMSC are learned through contacts, connections, and conditioning. A symbolic-psychological-emotional undercurrent would most likely be present. (p. 79)

UNCONSCIOUS, TYPE 8: DREAM STATES & THE EXPRESSION OF IMAGINAL CONNECTIONS. Since *SM* do not go into sleep states and dream in any conventional sense, it would not appear that this form of unconscious expression would apply. However, imagination may play a role, although it is unclear whether this would allow *SM* to drift into an unconscious state of awareness, and whether they would have some of the experiential qualities of dreaming. A symbolic-psychological-emotional undercurrent may be present, depending on the nature of the imaginative experience. (pp. 79-80)

UNCONSCIOUS, TYPE 9: THE INNATE & NATURAL EXPRESSION OF SPIRITUS. This is also an unknown. Could *SM* have *spiritus* as an element in their personality? This would imply that individual *SM* would possess unique and individual qualities and characteristics from their beginning. Implicit in this question is whether God, the cosmos, or some other unknown universal energy has had a hand in *SM*'s existence. Since *SM* are not biologically based but an engineering-electronic marvel, the possibility that genetics has played a role in their individuality is eliminated and cannot explain a *spiritus* element in the personality. This leaves only the esoteric-spiritual realm to explain *spiritus*'s possible presence and influence on *SM*, and impact on individual differences among *SM*. If *SM* were to possess a *spiritus* element, no symbolic-psychological-emotional undercurrent would be present. (p. 80)

UNDERLYING MOTIVATIONS. *SM* finding "balance" may simply be an excuse to avoid making tough decisions or an example of being influenced by their "better angels." Motivational factors of these kinds may include: fear (*SM* avoiding taking the necessary risks to resolve an issue), self-indulgence (casting off civil behavior as an expression of narcissism and pleasure-seeking), or something of significance (fulfilling a commitment to themselves or others). (p. 149)

UNDERSTANDING. *SM* being savvy, smart, and possibly shrewd in dealing with ideas, people, other *SM*, and institutions. (p. 192)

UNINFORMED. *SM* relying on inaccurate and unreliable information. (p. 201)

UNIQUENESS OF PCs. *SM*'s *PCs* will not necessarily have all the elements of the personality, and some may have only a single element. (p. 218)

UNNECESSARY CAUTION. *SM*'s fear of making mistakes and the consequences. (p. 183)

UNRESOLVED PERSONALITY CORRELATES. *SM* reveal a lack of development and integration among *PI, II, CI, TI, pathos*, and *directives*. (p. 209)

VALIDATION. Providing affirmations without an exaggeration of

SM's achievements. (p. 25)

VALUE CLARIFICATION. *SM* detailing their wants and needs and coming to terms with contradictions and inconsistencies in their life and lifestyle. (pp. 138-140)

VICARIOUS LIVING AND SPECTATOR SPORTS. Experiences through observing others, which can serve as a distraction from the stresses of modern life but also become an obsession. (p. 146)

VICTIMIZATION BY INSTITUTIONS. Societal institutions and corporations display cynical and hostile views, causing *SM* to adopt negative and maladaptive attitudes. (p. 129)

VICTIMIZATION BY OTHERS. Lacking empathy and morals, people and other *SM* have a strong influence on *SM* who then take on their exploitative traits. (p. 129)

VINDICTIVE. When *SM* believe they have been "wronged" and obsess on getting even. (p. 57)

VISUAL-SPATIAL MEMORY. *SM*'s ability to recall shapes, sizes, angles, locations, and movements. (p. 169)

VULNERABILITIES (see **SENSITIVITIES AND VULNERABILITIES**)

WANTS. *SM*'s desires ranging from mild attractions to full-blown cravings that can expand and evolve into compulsions that cause extreme behaviors. (p. 28)

WARMTH. *SM*'s emotional state that radiates friendliness, closeness, and familiarity. (p. 25)

WEIGHTING & RANKING OF VALUES. How strongly *SM* feel

about their wants and needs, reflected in assessing values' importance and significance. (p. 139)

WILLINGNESS TO ACT. *SM*'s values require them to take steps that involve risking their reputation, career, or lifestyle. (p. 138)

WISDOM. *SM*'s integration and synthesis of what is known, and the ability to apply knowledge in a fruitful and bountiful way to gain something positive. (p. 126)

WITHDRAWAL (see **CENSORING OF TFA**). A form of censoring that involves *SM* becoming socially inhibited, primarily due to generalized anxiety. (p. 145)

ABOUT THE AUTHOR

MARIANO GALLO is the author of *Clearmind: The Disentanglement of the Psyche*, a framework for understanding one's personality and thinking. He received a Ph.D. in clinical psychology from Miami University (Ohio) upon completion of a clinical residency and dissertation study at Northwestern University Medical School. Previous degrees include an M.A. in clinical psychology from Roosevelt University and a B.S. in psychology from Loyola University. He is a past fellow of the National Institute of Mental Health (NIMH) and a life member of the American Psychological Association (APA). He is a co-founder of Consultants in Applied Psychology (CAP) and a licensed clinical psychologist in Illinois (active) and California (inactive). His primary focus is the psychology of personality and applying it to life events and experiences. Get in touch with Mariano at consultantsinappliedpsychology@gmail.com.

Made in United States
Orlando, FL
04 November 2026

90205149R00196